Semiconductor Fundamentals

Robert F. Coughlin
and
Frederick F. Driscoll Jr.

Wentworth Institute

PRENTICE-HALL, INC. *Englewood Cliffs, New Jersey*

Library of Congress Cataloging in Publication Data

COUGHLIN, ROBERT F date
  Semiconductor fundamentals.

  Bibliography: p. 307
  1. Semiconductors. I. Driscoll, Frederick F.,
date joint author. II. Title.
TK7871.85.C65   621.3815'2   75-2314
ISBN 0-13-806406-7

Printed in the United States of America

PRENTICE-HALL INTERNATIONAL, INC., *London*
PRENTICE-HALL OF AUSTRALIA, PTY. LTD., *Sydney*
PRENTICE-HALL OF CANADA, LTD., *Toronto*
PRENTICE-HALL OF INDIA PRIVATE LIMITED, *New Delhi*
PRENTICE-HALL OF JAPAN, INC., *Tokyo*
PRENTICE-HALL OF SOUTHEAST ASIA (PTE.) LTD., *Singapore*

To the memory of *John L. Asinari,*
a student who bettered the places he passed
and the people he met

Contents

iv

15 Silicon-Controlled Rectifiers and Triacs   258

16 Frequency, Power, and Temperature Limitations   272

Preface

The purpose of this text is to present a direct approach to the understanding and use of semiconductor devices. It has been written at an introductory level for electronic and electrical technology students.

Chapters 1 through 8 lay a foundation for further study of semiconductor devices by concentration on the electrical behavior of *pn* junctions in diodes and bipolar junction transistors. Emphasis is placed on the common-emitter circuit because of its importance and wide usage. Circuits are studied by separating dc analysis from ac analysis. This simplifies understanding and follows actual practice, where one first checks dc voltages to see if the device is conditioned to operate and then injects an ac signal to begin operation.

Chapters 8 to 11 build on this foundation. Practical applications of bipolar junction transistors in amplifier circuits are studied. Chapters 12 and 13 are concerned with field-effect transistors. Extra attention is given to the complementary metal-oxide semiconductor (CMOS) family of devices, and applications are selected from the field of digital logic circuits.

Semiconductor devices such as unijunction transistors are introduced in Chapter 14 to show how they are used in timing and oscillator circuits. They are then employed in Chapter 15 to turn on power-control devices such as the triac and SCR. Frequency, power, and temperature limitations are discussed in Chapter 16. Power supplies are presented in Chapter 17, although this subject may be covered, at the user's discretion, any time after the material in Chapters 1 through 8 has been assimilated.

Many examples have been included to show how the principles are applied in the solution of practical problems. Problems at the end of each chapter have been carefully matched to the examples to ensure greater understanding, through comparison.

The authors gratefully acknowledge the advice and guidance furnished by Dean Charles M. Thomson. We thank Mary Hatfield for her cheerful and skilful manuscript preparation.

ROBERT F. COUGHLIN
FREDERICK F. DRISCOLL

Boston, Massachusetts

1

Semiconductor Material

1-0 Introduction

Less than two decades ago, the well-known vacuum tube, in a variety of shapes and sizes, performed just about all the tasks associated with the rather minor engineering field of electronics. Electronic applications were then concerned mostly with radio communications. In our current world we are aware of the enormous influence of electronics, which touches nearly every aspect of our lives. Modern industrial electronic periodicals presently devote only about 3 percent of their space to communications equipment. Furthermore, it is growing difficult to find an advertisement for vacuum tubes except in sections listing surplus equipment of service-oriented electronics magazines.

It was semiconductor devices that toppled the vacuum tube in all but a few areas of electronics, and even these areas are presently under vigorous assault. Semiconductors made inevitable a fantastic expansion of electronics into the fields of computer technology, industrial controls, medical technology, space exploration, life sciences, chemistry, physics, politics, and, regrettably, war technology. Semiconductor devices range in size from tiny simple diodes to large power-controlling silicon-controlled rectifiers. They include extremely complex arrays of transistors fabricated on a single integrated circuit chip. Semiconductor chips containing all the electronics for portable calculators, watches, radios, hearing aids, pacemakers, and a variety of instrumentation and control systems are now available.

To understand these devices, this chapter introduces the physics of semi-

1

conductor materials. Specifically we will study atoms, ions, and the differences between semiconductor materials.

1-1 The Atom

1-1.1 Atomic Models. A model is a mathematical or physical aid that helps in the understanding of how something works. Three fundamental properties of an *atom* are shown in Fig. 1-1. The model used is that of a hydrogen atom.

1. The heavily lined inner circle is the *nucleus*. This contains most of the atom's mass and all its positive charges.
2. The particle traveling in an orbit around the nucleus is an *electron*. The electron has a (negative) charge of -1.
3. The number of positive charges in the nucleus equals the number of electrons in orbit; or, to put it another way, the atom is electrically neutral.

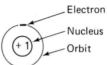

Figure 1-1 Hydrogen atomic model.

1-1.2 Carbon Atomic Model. The carbon atom is introduced now to illustrate additional properties of the atom. The carbon atomic model in Fig. 1-2(a) introduces four new ideas.

1. No two electrons in the same atom can ever occupy the same orbit.
2. An electron can occupy only certain orbits.

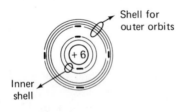

(a) Carbon atomic model

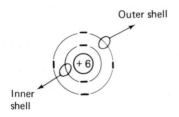

(b) Simplified model

Figure 1-2 Carbon atomic models.

3. From chemistry definitions, the orbits are grouped into *shells*. For example, of the six electrons in a carbon atom, the two inner electrons are grouped together in what is known as the *inner* shell. The remaining four electrons are grouped into an *outer* shell.
4. An atom of a particular element can be identified by an atomic number equal to the number of positive charges in its nucleus. Carbon's atomic number is six; hydrogen's atomic number is one.

It is easier to draw an atomic model by representing a shell as a single ring. As shown in Fig. 1-2(b), this technique quickly distinguishes one shell from another. Remember, however, that the electrons within a shell actually travel in different orbits.

1-1.3 Bohr Atomic Model. In 1913 Niels Bohr proposed a model to approximate the atomic makeup of all the natural elements, ranging from the lightest element, hydrogen, to the heaviest, uranium. This model, Fig. 1-3,

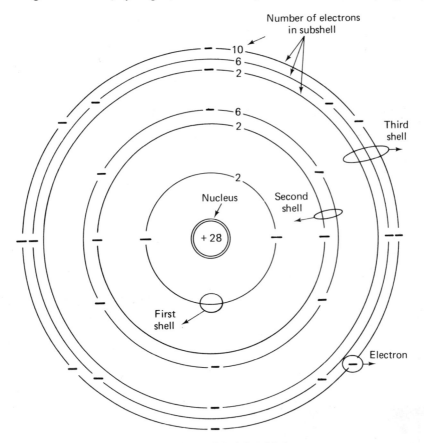

Figure 1-3 Bohr model of the nickel atom.

shows that each shell may be divided into *subshells*. Each subshell can contain only a certain number of electrons. Moving out from the nucleus, a subshell, of each shell, can contain only four electrons more than the preceding subshell. An understanding of semiconductor operation involves an examination of an atom containing the first three shells. This is because silicon, with an atomic number of 14, is the most common semiconductor material, and it needs only three shells to contain all its electrons.

1-1.4 Valence Electrons. The number of electrons in the outermost shell of an atom is of particular importance because these electrons determine the electrical properties of the material. A material that has only a few (one or two) electrons in the outermost shell will require very little energy to free these electrons from the atom. These freed electrons will wander through the material and allow the material to conduct current easily. If, on the other hand, a material's atoms have an outer shell that is completely filled, large amounts of energy will be required to free electrons. These materials are poor conductors of electric current and are therefore classified as insulators. A semiconductor is a material that has its outermost shell partially filled. Normally at room temperature a pure-semiconductor material is neither a good conductor nor a good insulator. Since electrons in the outermost shell are so important, they have a special name, *valence electrons*. Materials most commonly used in the manufacture of semiconductors are germanium and silicon, both of which have four valence electrons. Of these two materials, silicon is the most widely used.

1-1.5 Silicon Atom. As shown in Fig. 1-4(a), the silicon atom has three shells. Those electrons occupying orbits in the outer shell are identified by the term "valence electrons." Since the chemical and electrical behavior of an atom depend primarily on its valence electrons, it is simpler to represent the silicon atom by the model in Fig. 1-4(b). Here only valence electrons are shown, together with their associated positive charges in the nucleus. One other term should be learned at this point—*tetravalent*. Tetravalent identifies a group of elements that contain four electrons in the outer shell. Carbon, germanium, silicon, tin, and lead are all tetravalent atoms.

1-2 Doping Atoms

Before studying what the doping process is and why it is needed, we need to look at two other groups of atoms, trivalent and pentavalent.

1-2.1 Trivalent Atom. The *trivalent atom* (tri = three) has three valence electrons in its outer shell. Boron, aluminum, gallium, and indium atoms

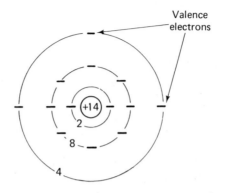

(a) Model of silicon atom

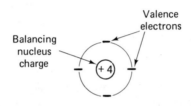

Figure 1-4 Silicon atomic models.

(b) Simplified model

are all trivalent. Each can be modeled, as shown in Fig. 1-5(a), by three electrons and their associated three positive nuclear charges. It is important to note that the atom is electrically neutral, with three minus charges exactly balanced by three positive charges.

1-2.2 Pentavalent Atom. The *pentavalent atom* (penta = five) contains five valence electrons in its outer shell. Phosphorus, arsenic, and antimony atoms are pentavalent. Each can be modeled as shown in Fig. 1-5(b) by five positive nuclear charges. Like trivalent and tetravalent atoms, the pentavalent atom is electrically neutral.

(a) Trivalent atom

(b) Pentavalent atom

Figure 1-5 Doping atoms.

1-3 Ions

An atom is said to be *ionized* or becomes an *ion* when it gains or loses one or more electrons in its outer shell. Atoms of a solid are fixed in their atomic structure and cannot move at temperatures below their melting point. So, when an atom is *ionized*, only an electron or electrons are removed or added; the resulting ion remains fixed in position.

1-3.1 Negative Ion. For our example of creating a negative ion, let's consider Fig. 1-6(a) a neutral trivalent atom. It is possible under circumstances discussed in Section 1-6.1 for a trivalent atom to gain an extra electron. When this occurs, as in Fig. 1-6(b), our valence electrons are offset by only three positive charges in the nucleus. The trivalent atom has become a negative ion and has a net negative charge of -1. In semiconductor technology it is much more common for a trivalent atom to *accept* an extra valence electron rather than lose or *donate* a valence electron. For this reason, trivalent atoms are often called *acceptor atoms.*

1-3.2 Positive Ion. The pentavalent atom of Fig. 1-6(c) will be used to demonstrate the idea of a positive ion. As will be shown in Section 1-6.2, a pentavalent atom will often lose or *donate* one of its valence electrons. Therefore, the pentavalent atom becomes a positive ion with a net charge of $+1$. Figure 1-6(d) shows the five positive charges of the nucleus offset by only four negative valence electrons. Since a pentavalent atom loses or donates a valence electron, it is often called a *donor atom.*

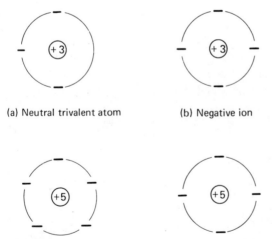

(a) Neutral trivalent atom

(b) Negative ion

(c) Neutral pentavalent atom

(d) Positive ion

Figure 1-6 Examples of negative and positive ions.

1-4 Pure Semiconductor Material

1-4.1 Semiconductor Resistance. A semiconductor is a material whose resistance lies between the low resistance of a conductor and the high resistance of an insulator. Table 1-1 gives typical resistance values for comparison.

To learn how a semiconductor conducts current, we must see what happens when semiconductor atoms are brought together.

Table 1-1 COMPARISON OF RESISTANCE

Material	Classification	Approximate Resistance (Ω/cm^3)
Silver	Conductor	10^{-5}
Pure silicon	Semiconductor	50×10^3
Pure germanium	Semiconductor	50
Mica	Insulator	10^{12}

1-4.2 Pure Semiconductor. One way to construct a semiconductor crystal is to melt extremely pure silicon and then slowly cool it around a silicon seed crystal. As shown by the two-dimensional model in Fig. 1-7, the silicon

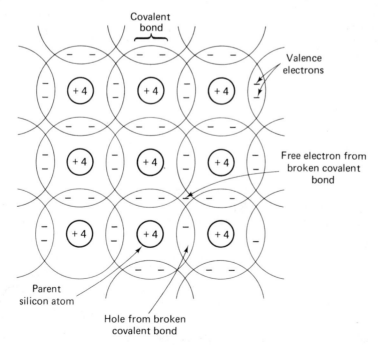

Figure 1-7 Pure semiconductor crystal.

atoms become arranged, like the seed crystal, in a uniform pattern called a *cubic lattice*.

Note that each of the four valence electrons of a single atom can share an outer orbit with an electron from a neighboring atom. So each atom has the equivalent of eight electrons, or four electron pairs, in its outer orbit. This arrangement is extremely stable, both electrically and chemically. The electron pair is also called a *covalent* bond. Covalent means shared valence electrons and bond means holding atoms together.

1-4.3 Rupture of a Covalent Bond. At a temperature of absolute zero ($-273°$ centigrade), electrons orbit relatively close to the nucleus. As energy is absorbed by the silicon crystal, electrons orbit faster and at greater distances from the nucleus. Some atoms eventually absorb enough energy to break one of their covalent bonds. As shown in Figure 1-7, this results in one valence electron escaping from its parent atom. Figure 1-7 is drawn primarily to illustrate the crystal makeup of tetravalent atoms. Actually the crystal is mostly empty space, so the escaped electron is free to wander through the crystal as a free negative charge. It will probably never return to its parent atom.

1-4.4 Hole Motion. As noted, when a covalent bond is broken, an electron is freed to wander through the empty space of the semiconductor lattice. The escaping electron creats an absence of one electron in the structure of its parent atom. This absence is called a *hole*. The parent atom, normally electrically neutral, now has an excess of one positive charge and is therefore a positive ion with a net charge of $+1$. We can model or represent the hole by a single positive charge, equal in magnitude but opposite in sign to the charge of an electron.

Since atoms in a solid cannot move, the positive ion that was just created cannot move. However, a valence electron from a neighboring atom can move into the hole. Now once again, the parent atom becomes electrically neutral but the neighboring atom becomes a positive ion, as shown in Fig. 1-8. The hole has, in effect, moved from the parent atom to the neighboring atom. Thus a hole or positive charge is considered to move.

1-4.5 Hole-Motion Analogy. For an analogy of hole motion, consider Fig. 1-9. The four cars represent valence electrons. Car 1 gains enough energy and moves past the stop sign. This is analogous to breaking a covalent bond. When car 1 moves forward a space (or hole) is created, Fig. 1-9(b). Now car 2 is able to move forward, leaving a space where it had been, Fig. 1-9(c). When car 3 moves forward the space moves backward, Fig. 1-9(d); and when car 4 moves forward, the space once again moves backward, Fig.

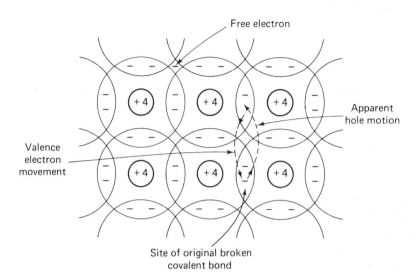

Figure 1-8 Hole motion.

1-9(e). To an observer the event can be viewed as either (1) the movement of cars one at a time from right to left or (2) the movement of a space from left to right. The second situation would be described as hole motion.

Let us now turn our attention back to car 1. It has two possible routes: (1) with enough energy it could go up the ramp and be free to travel at its own will. This is analogous to a covalent bond being broken and the freed electron being able to wander through the empty space in the semiconductor material. (2) Without sufficient energy it takes the downtown route and becomes enmeshed in traffic; it then moves forward one space at a time. This is analogous to hole motion that occurs after the covalent bond is broken.

1-5 Current Magnitude in a Pure Semiconductor

Current flow is the net motion of charges in one direction. In a semiconductor, current flow may consist of either positive holes or free negative electrons. But in pure silicon there is, at room temperature, only about one broken bond for every 3 million million silicon atoms. In pure germanium (another semiconductor material) about one broken bond occurs for every 2 billion germanium atoms. This means that the resistance of germanium is roughly 1000 times less than that of silicon, but both materials have so few available charge carriers at room temperatures that they can conduct very

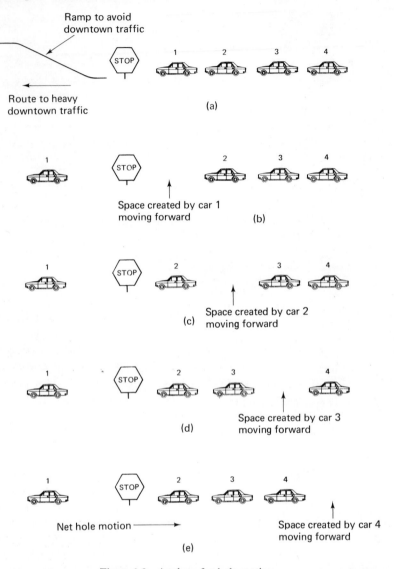

Figure 1-9 Analogy for hole motion.

little current. Therefore, to realize currents in the order of milliamperes or amperes, the pure semiconductor material must be modified. This modification is accomplished by adding more free charge carriers, either holes or electrons. The procedure that accomplishes this is called *doping*. (*Note:* Doping a pure semiconductor material with either holes or electrons lowers the resistance of the material, thereby allowing increased current values.)

1-6 Doping

As previously stated, doping is a manufacturing process that adds free charge carriers (holes or electrons) into pure semiconductor material to lower its resistance. There are two basic doping processes. One process is the addition of positive charges or holes. The other process is the addition of negative charges or electrons. Typically doping reduces resistance of pure semiconductor material by a factor of 100,000. We shall now discuss how free charges are added by doping.

1-6.1 *p*-type Semiconductor Material. Recall from Section 1-3.1 that an acceptor atom has three valence electrons. It is seen from Fig. 1-10(a) that, when an acceptor atom is added by doping to pure silicon, only three covalent bonds are formed. This leaves a hole in the acceptor atom's outer shell that can be occupied by a valence electron. Hole motion can occur just as described in Section 1-4.4. When a valence electron from a neighboring silicon atom moves into the hole, the acceptor atom becomes a negative ion with a net local charge of -1. The hole, in effect, is free to move as a result of a valence electron moving to fill a hole. The hole is then referred to as a *current carrier*.

Roughly one acceptor atom, and consequently one hole, is added for every 10 million silicon atoms. Compare this quantity with the one electron and one hole generated by one broken bond for every 3×10^{12} silicon atoms. Thus for every free electron there are about 300,000 holes. Clearly holes are in the majority, so holes are called *majority current carriers* in a *p*-type semiconductor. Therefore, electrons are called *minority current carriers* in *p*-type semiconductor material.

1-6.2 *n*-type Semiconductor Material. In the manufacturing process of an *n*-type semiconductor, roughly one donor atom (a pentavalent atom) is added by the doping process to every 10 million silicon atoms [see Fig. 1-10(b)]. The donor atom (see Section 1-3.2) forms covalent bonds with four neighboring silicon atoms. At room temperature, the donor atom's fifth electron has more than enough energy to escape and become a free current carrier. Without doping there was only one hole and one electron (created by a broken covalent bond) for every 3×10^{12} silicon atoms. Therefore, doping has added about 300,000 free electrons for every free hole. Thus electrons are the *majority current carriers* in this *n*-type semiconductor material and holes are *minority current carriers*.

In the remainder of this text the term "semiconductor" will mean germanium or silicon material that has already been properly doped with either acceptor or donor atoms.

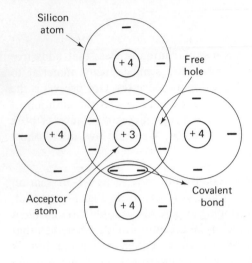

(a) p-type Semiconductor

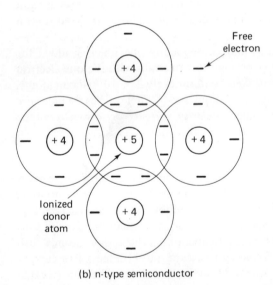

(b) n-type semiconductor

Figure 1-10 Doping with acceptor atoms in (a) and donor atoms
in (b) gives *p*-type and *n*-type
semiconductor material, respectively.

1-7 Diffusion and Drift Currents

1-7.1 Thermal Motion of Charge Carriers. There are two methods by
which charge carriers move to constitute a current within a semiconductor.
Both depend on the idea of thermal motion. For example, once an electron
becomes free it moves through the mostly empty space of the lattice with an
initial velocity and direction. Visualize the semiconductor atoms as vibrating

so that a free electron colliding with an atom will rebound with more or less speed and probably change direction. The same analogy applies to free holes. Thus free holes and electrons move in random fashion, in different directions and at different speeds. This motion is called *thermal motion*.

1-7.2 Diffusion Current. In Fig. 1-11(a) one end of a semiconductor is heated to create an excess of holes and electrons (by breaking covalent bonds). The freed electrons and holes will have thermal motion (as described in Section 1-7.1), and therefore some must eventually move by chance to the right, as shown in Fig. 1-11(b). After sufficient time, holes and electrons will be distributed uniformly throughout the sample, as shown in Fig. 1-11(c). Since charge carriers have moved clearly in one direction, current flow has taken place. Diffusion current is analogous to the diffusion of perfume molecules. If perfume is spilled in one corner of a room, eventually the perfume molecules diffuse evenly throughout the room.

1-7.3 Drift Current Within a Semiconductor. When a voltage is connected across a semiconductor, an electric field is set up to attract holes toward the minus connection and electrons toward the plus connection. The effect of the electric field on the free charge carriers is not very strong because they are screened by atoms in the crystal lattice. The applied electric field thus changes normal thermal motion slightly, and the net effect is a *drift* of free holes in the direction of the electric field. In Fig. 1-12(a) a free hole is

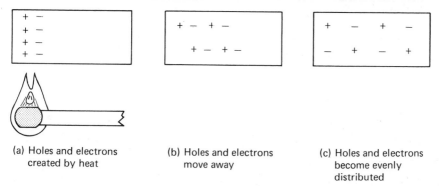

(a) Holes and electrons created by heat

(b) Holes and electrons move away

(c) Holes and electrons become evenly distributed

Figure 1-11 Diffusion current in a semiconductor.

modeled by a plus sign. For simplicity the lattice is not shown. Direction of the electric field is shown by the arrow and, by definition, the arrow points in the direction that the field would move a positive charge (hole).

Figure 1-12(a) represents the thermal motion of a single hole in the absence of an electric field. Figure 1-12(b) shows how the same motion would be modified over the same time interval when an electric field is added. The

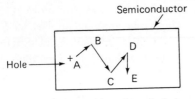

(a) Thermal motion of a hole

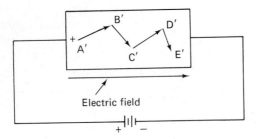

(b) Hole motion is changed by
an electric field

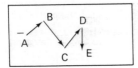

(c) Thermal motion of a
free electron

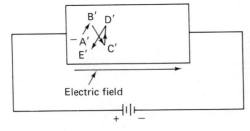

(d) Electron motion is modified
by an electric field

Figure 1-12 Drift currents within a semiconductor.

net effect of the electric field is to drift the hole to the right, as seen by comparing locations E and E'. Compare Fig. 1-12(c) and (d) to see that the electric field results in a net drift to the left for a free electron.

It could be argued that only electrons drift to the left in Fig. 1-12. That is, free electrons drift left and holes drift right. Remember, holes drifting right is equivalent to valence electrons (not free electrons) drifting left.

1-8 External Current Flow

A metal pad of material such as aluminum or gold is fused onto the semiconductor to form a contact. Wires are then bonded to the contacts so that the semiconductor may be connected into a circuit. The metal–semiconductor contacts are designed as *ohmic* contacts. It is the job of the ohmic contact to allow the semiconductor that conducts both holes and electrons to work in series with a metal wire that conducts only electrons.

In Fig. 1-13 the battery sets up an electric field that drifts holes right and electrons left. In the external circuit electrons are conducted through the wire in a direction away from the (−) battery terminal and toward the (+) battery terminal. There is one wire electron for each free semiconductor electron and also for each free semiconductor hole that reaches an ohmic contact. Conventional current flow is also shown in Fig. 1-13 as being opposite to electron flow in the external circuit. Armed with these fundamental properties of atoms, ions, and semiconductor materials, we proceed to Chapter 2, where we examine what happens when *p*-type and *n*-type materials are joined.

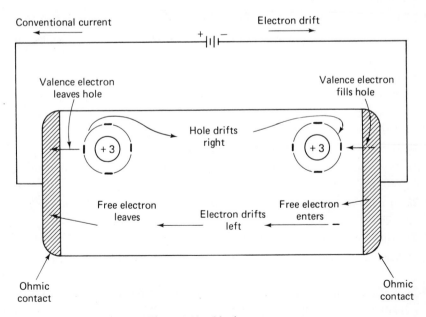

Figure 1-13 Ohmic contacts.

Problems

1-1 What is that part of an atom called that contains most of the weight of the atom and all its positive charge?

1-2 What is the term for the negative particle that travels in an orbit around the nucleus?

1-3 What are orbits that surround the nucleus grouped into?

1-4 What is the atomic number equal to?

1-5 What term is used to identify electrons in the outermost shell?

1-6 Is an atom with its outermost shell completely filled a good conductor of electric current?

1-7 Define tetravalalent.

1-8 Define trivalent.

1-9 Define pentavalent.

1-10 Semiconductors belong to what group of elements?

1-11 What is the group of elements that contains boron, aluminum, gallium, and indium?

1-12 What is the name given to an atom that gains or loses one or more electrons?

1-13 What is an atom called when it gains an electron?

1-14 Define donor atom.

1-15 Is the resistance of a semiconductor material lower than that of an insulator?

1-16 What name is used to describe the uniform pattern of a pure semiconductor material?

1-17 What is the absence of an electron called?

1-18 What is created when a covalent bond is broken?

1-19 Does doping a pure semiconductor material increase the resistance of the material?

1-20 Define *p*-type semiconductor material.

1-21 In a *p*-type semiconductor, what are (a) majority carriers; (b) minority carriers?

1-22 In an *n*-type semiconductor material, what are (a) majority carriers; (b) minority carriers?

1-23 What are the two methods by which charge carriers move in a semiconductor?

1-24 What is the name given to the type of current produced by freed electrons and holes as a result of thermal motion?

1-25 When freed electrons and holes move because of an electric field, what is the type of current called?

2

pn Junction

2-0 Introduction

When a sample of *p*-type semiconductor material is joined to a sample of *n*-type semiconductor material, the interface between them is called a *pn junction*. It is the electrical characteristics of the junction rather than the semiconductor sample material that determines the electrical behavior of the entire structure. In fact, if we understand how the *pn* junction works, we will have taken the first step toward understanding the operation of all semiconductor devices. We begin by examining what happens when a *pn* junction is created.

2-1 *pn* Junction

2-1.1 Space-Charge Region. To explain what happens at the junction or interface between *n*-type and *p*-type semiconductor materials, refer to Fig. 2-1(a). Free electrons and holes are shown as − and + symbols, respectively. Locked (cannot move) positive donor atoms and locked negative acceptor ions are shown as circled + and − symbols, respectively ($\oplus$ and $\ominus$).

Electrons from the *n*-type material near the junction diffuse or wander across the junction. These electrons fill the holes in the *p*-type material adjacent to the junction. As a result of electrons leaving the *n* material, donor ions are created on the *n* side of the junction. When these electrons fill holes on the *p* side of the junction, acceptor ions are produced. These ions occur only

17

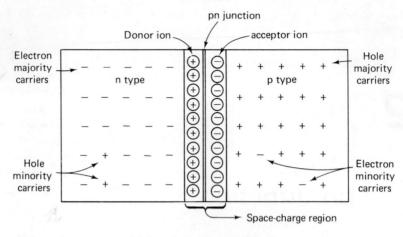

(a) Formation of a pn junction

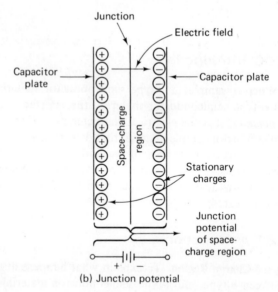

(b) Junction potential

Figure 2-1 Space-charge-region capacitance and junction potential.

near the junction, as shown in Fig. 2-1(a). A wall of stationary positive ions is aligned with a wall of stationary negative ions along the *n* and *p* sides of the junction, respectively. The space occupied between the ion walls is called a *space-charge region*.

Whenever there exists a positive charge with respect to a negative charge, a voltage or potential difference is set up between the charges. A potential difference, called the *junction potential,* is therefore established across the *pn* junction.

2-1.2 Junction Potential and Capacitance. When a potential difference is established across a charged capacitor, an electric field exists between plates of the capacitor, as shown in Fig. 2-1(b). An electric field must also be established across the space-charge region of a *pn* junction because of the junction potential. When we compare a charged capacitor and the *pn* junction we conclude that the space-charge region has properties of a capacitor.

Consider the properties of a charged capacitor. If a positive charge is placed between the plates of a charged capacitor, the positive charge will be repelled by the positive plate and attracted toward the negative plate. A negative charge placed between the charged plate would be repelled by the negative plate and attracted toward the positive plate.

With these properties in mind we shall now investigate how the junction potential affects movement of free charge carriers (both holes and electrons) within the space-charge region.

2-1.3 Action of the Junction Potential. Free electrons and holes have a wide range of energy. This idea is simplified if we relate electron energy to speed. Visualize free electrons as moving randomly and colliding with atoms. Some will rebound faster (more energy) and some will rebound slower (less energy). None of the free charge carriers in the *n*-type or *p*-type material is affected by the electric field that exists at the junction. However, because of their thermal motion, holes and electrons will eventually diffuse or drift into the space-charge region. What happens to these charge carriers in the space-charge region depends on whether the carriers are majority carriers or minority carriers. If majority carriers, we must also take into consideration their velocity when they enter the region.

As shown in Fig. 2-2, the junction potential acts as a *potential barrier* that tends to prevent majority carriers from crossing the junction. In *n* material the majority carriers are electrons and most of the electrons that try to cross the junction are repelled by the space-charge regions electric field. The exceptions are a few high-energy electrons that are able to cross the junction. Similarly, majority holes from the *p* side that enter the space-charge region are returned to the *p* side, with the exception of a few with sufficient energy to cross the junction. The electric field arrow in Fig. 2-2 points in the direction a positive charge would move.

Minority carriers (holes on the *n* side and electrons on the *p* side) are aided by the junction potential. In fact, their passage through the space-charge region is actually accelerated across the junction by the electric field. If a hole from the *n* side wanders into the space-charge region, it is strongly

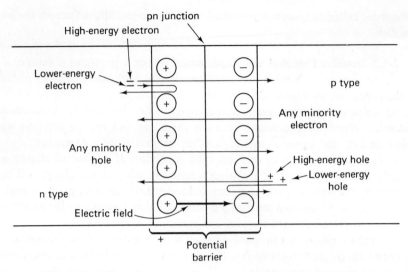

Figure 2-2 Effect of junction potential barrier on holes and electrons.

attracted by the negative ions on the *p* side. Similarly, an electron from the *p*-side that wanders into the region is accelerated across the junction by the electric fields.

2-1.4 *pn*-Junction Equilibrium. A dynamic equilibrium exists in the *n*- and *p*-type regions of Fig. 2-2. By dynamic equilibrium we mean that the holes and electrons are in constant motion. However, there is no net current flow because the probability of finding a hole or an electron moving in one direction is the same as that of finding another hole or electron moving in the opposite direction. For our sample, high-energy electrons on the *n* side can cross the junction to become minority electrons on the *p* side. However, an equal number of *p*-side minority electrons will wander into and be accelerated back across the space-charge region. The same mechanism applies in reverse to holes. This action ensures not only that the entire sample remains electrically neutral but also the *n* side and *p* side. Also, there is no net motion of charges in one direction, and therefore there is no current flow.

2-1.5 Depletion Region. We have seen that free charges cannot remain long in the space-charge region. Minority carriers are accelerated across it and majority carriers are swept back. For this region the space-charge region is also called the *depletion region*. That is, it is a region depleted of charge carriers. We reason that a region with no charge carriers must be one of high electrical resistance. This conclusion will be used in Chapter 3 to find how a device made with a *pn* junction will behave in a circuit. (*Note:* The region near the junction of *p* and *n* materials is referred to by many names: space-charge region, potential barrier, and depletion region, but the most common term is "*pn* junction.")

2-2 Biasing a *pn* Junction

In Section 2-1.1 we have shown that positive ions form on the *n* side of the junction while negative ions form on the *p* side. This difference in charge is a potential difference or voltage. The value of this voltage depends on the semiconductor material used. Until now we have studied the *pn* junction with no outside energy being applied (other than heat and light). We now wish to examine what happens when an external battery is connected across the *pn* junction. This is called *biasing* the *pn* junction. There are two ways to bias a *pn* junction, by forward bias and by reverse bias.

2-2.1 Forward Bias of a *pn* Junction. When the positive terminal of a battery is connected to the *p* side and the negative terminal is connected to the *n* side, the *pn* junction will be *forward-biased* (see Fig. 2-3). Forward-bias voltage, V_F, causes a decrease in the junction potential across the space-charge region. With this decrease the number of majority carriers to cross the junction increases because less energy is required.

Holes from the *p* side cross the junction, pass the positive ion wall, and

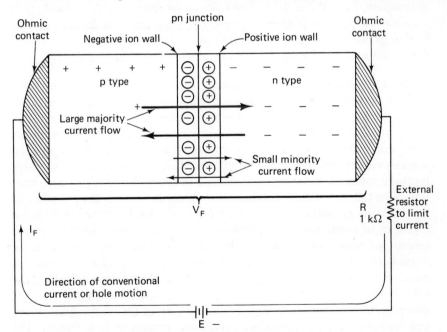

Figure 2-3 Forward bias of a *pn* junction.

begin to diffuse through the *n* material. These holes will eventually combine with free electrons in the *n* side. Similarly, free electrons cross the junction, pass the negative ion wall, and begin to diffuse through the *p* material. These free electrons will combine with holes on the *p* side. The reason there is always a large majority of holes on the *p* side and electrons on the *n* side is because the battery, along with the ohmic contacts, replenishes them (see Section 1-8). Figure 2-3 also shows that minority carriers that wander into the space-charge region are swept across the junction.

Unlike the example of a *pn* junction with no external supply described in Section 2-2, the number of majority carriers crossing the junction is much greater, 1000 or even 1 million times greater. Therefore, there is a net flow of charges in one direction and this net motion is current flow. External resistor *R* is connected in series with both battery and *pn* junction to limit the value of current I_F so that no component is destroyed.

Conventional current flow is from positive to negative. In other words, it is in the direction a positive charge or hole moves. (Electron motion is in the opposite direction.) Current I_F signifies forward current because the junction is forward-biased. It is shown as conventional current in Fig. 2-3 to agree with the diode's symbol, which will be shown in the next section.

The value of V_F depends on whether the material used in the manufacture of the *pn* junction is germanium or silicon. If $I_F = 1$ mA and the material is germanium, $V_F = 0.2$ to 0.3 V. For a silicon *pn* junction, if $I_F = 1$ mA, then $V_F = 0.6$ to 0.7 V. A picture or graph that shows how the forward current I_F changes as V_F is varied is called *forward current–voltage characteristics*. In Chapter 3 we shall discuss these characteristics and how to use them.

2-2.2 Reverse Bias of a *pn* Junction. To reverse-bias a *pn* junction, connect the positive terminal of a battery to the *n* side and the negative terminal to the *p* side, as in Fig. 2-4. Reverse-bias voltage, V_R, causes an increase in the width of the space-charge region and an increase in the junction potential. With this increase the number of majority carriers to cross the junction decreases to zero because none has sufficient energy to overcome the increased junction potential. However, this increase in space-charge region and junction potential is extremely conductive to minority current flow. Remember that minority current is electrons from the *p* side that enter the space-charge region to be swept across the junction to the *n* side. Also, minority current consists of holes from the *n* side that enter the space-charge region to be swept across the junction to the *p* side. Once again we have a net motion in one direction and therefore current flow. Unlike the forward bias situation, the minority current with a reverse bias is very small but greater than the zero majority current. I_s, in Fig. 2-4, is the symbol for *reverse bias current* and is shown in the direction of conventional current or hole motion. The subscript *s* stands for saturation because all available minority carriers are being conducted. The number of minority carriers depends on the number of

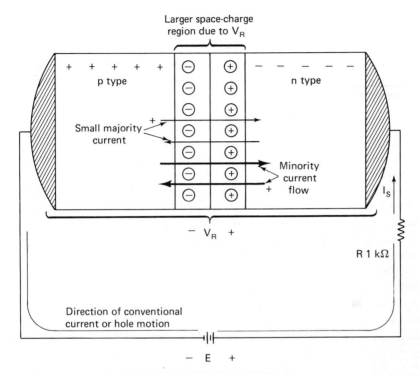

Figure 2-4   Reverse bias of a *pn* junction.

broken covalent bonds, which is temperature-dependent and does not depend on magnitude of reverse bias.

To compare values, I_F is in the order of milliamperes or amperes, whereas I_s is in the order of microamperes for a germanium *pn* junction and nano-amperes for silicon. In summary, a reverse-biased *pn* junction is characterized by a small reverse current I_s, which is practically independent of V_R. A forward-biased *pn* junction is characterized by a large forward current, I_F, and a small forward voltage, V_F.

2-3 Breakdown of a *pn* Junction

All *pn* junctions have a limit of maximum reverse-bias voltage beyond which the current increases rapidly. This voltage is called *breakdown voltage*. There is no typical value of breakdown voltage, because nearly any value greater than 3 V can be purchased inexpensively. There are two mechanisms that cause a *pn* junction to break down.

2-3.1 Avalanche Breakdown. It was stated in Section 2-2.2 that when a *pn* junction is reverse-biased, a saturation current flows. Saturation current

I_s is the result of minority carriers crossing the junction, that is, electrons from the *p* side entering the space-charge region and being swept into the *n* side, along with holes from the *n* side crossing the junction to the *p* side. Now, if the reverse-bias voltage is large enough, some of these minority carriers will gain sufficient energy so that if they collide with an atom a covalent bond is broken. This broken bond produces a new hole and a new electron, each of which can gain enough energy from the reverse-bias voltage to cause another broken bond. As this process is repeated, the saturation current quickly increases because almost all the carriers—holes and electrons—that are produced contribute to the total current. This type of breakdown is *avalanche breakdown*.

2-3.2 Zener Breakdown. Two things happen when a *pn* junction is reverse-biased: (1) the width of the space-charge region increases, and (2) an extremely large electric field is developed. It is this large electric field that causes Zener breakdown. The electric field causes covalent bonds to rupture, which produces large quantities of holes and electrons. These holes and electrons cause the rapid increase in saturation current. In comparison, avalanche breakdown is the result of collisions of high-energy minority carriers with atoms to break the covalent bond. Zener breakdown is the result of a strong electric field at the junction to break the covalent bond.

Pn junctions that break down below 5 V are caused primarily by Zener breakdown. Above 8 V, *pn* junction breakdown is predominantly due to avalanche mechanism. Between 5 and 8 V, the breakdown mechanism depends upon the amount of doping atoms and their distribution at the junction.

2-4 Circuit Symbol

Another name for *pn* junction is *diode*. Thus a diode is a two-terminal device that conducts current easily in one direction but allows only a small saturation current when reverse-biased. The diode circuit symbol is shown in Fig. 2-5(b). The same symbol is used regardless of whether the diode is germanium or silicon. The names of the diode's terminals are *anode* (*p*-type material) and *cathode* (*n*-type material). The terms "anode" and "cathode" are a carryover from vacuum-tube terminology. The arrowhead of the diode symbol has three significant meanings:

1. It is in the direction of conventional current flow when the diode is forward-biased.
2. It represents the anode or *p*-type material.
3. It points toward the *n*-type material.

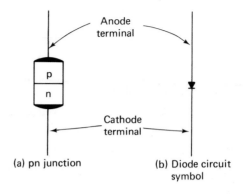

(a) pn junction   (b) Diode circuit
symbol

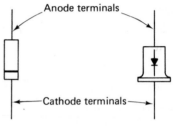

Figure 2-5 Diode symbol and
cases.   (c) Different types of diode cases

Figure 2-5(c) illustrates two types of cases that are used in the construction of semiconductor diodes. Note that the cathode end is easily identified
by (1) a band around the case, (2) the circuit symbol drawn on the case, or
(3) the terminal lead that has an eyelet for easy soldering (not shown).

Problems

2-1 Do donor or acceptor atoms move in a semiconductor to constitute current
flow?

2-2 What name signifies the region on either side of a *pn* junction?

2-3 Does the junction potential oppose or aid passage of majority carriers across
the junction?

2-4 Does the junction potential oppose or aid passage of minority carriers across
the junction?

2-5 Does an arrow that symbolizes an electric field point in the direction that a
free positive or negative charge would move?

2-6 Why is the space-charge region also called the depletion region?

2-7 How should a battery be connected to forward-bias a *pn* junction?

2-8 Does a forward bias increase or decrease the junction potential?

2-9 Why is a resistor connected in series with a forward-biased diode?

2-10 Does the arrowhead of a diode symbol point in the direction of conventional or electron current flow?

2-11 What is the value of forward-bias voltage V_F for a silicon diode?

2-12 The cathode of a diode connects to *p*- or *n*-type semiconductor material.

2-13 If the arrow end of a diode symbol is shown wired to a negative battery and the bar end is wired to a positive battery, is the diode reverse- or forward-biased?

2-14 Does the term "saturation current" apply to a forward- or a reverse-biased diode?

2-15 When the space-charge region's electric field is strong enough to rupture covalent bonds, is the resultant breakdown classified as Zener or avalanche breakdown?

2-16 Does the diode symbol's arrow point toward *p*-type material?

3

Diode Characteristics and Analysis

3-0 Introduction

To understand how any electronic device or system operates, we should know the relationships between voltage across it and current through it. This current–voltage relationship may be expressed as an equation such as *Ohm's law*, $V = IR$, or as a graph that is a plot of current through the device versus voltage across the device. The graph has the advantage of showing us immediately all possible values of current for corresponding values of voltage. Engineers and technicians use these graphs to obtain practical solutions to problems and thereby avoid unnecessary and complex equations. These graphs are referred to as the *I–V characteristic curves*. Most characteristic curves are plotted with the vertical or *y* axis as current and the horizontal or *x* axis as voltage. We shall show typical–characteristics for all devices introduced in this text. The diode can be either forward- or reverse-biased; therefore, it has both forward and reverse characteristics.

3-1 Current-Voltage Characteristics of a Semiconductor Diode

3-1.1 Forward Characteristics. In Chapter 2 we showed a forward-biased diode where a positive voltage is applied to its *p* side and a negative voltage to its *n* side. Under this condition a large majority current flows, and currents of either milliamperes or amperes can be realized. Figure 3-1(a) is a plot of

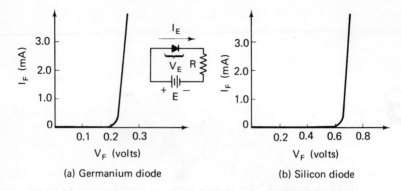

Figure 3-1 Typical forward characteristics of semiconductor diodes.

forward current, I_F, versus forward voltage, V_F, for a germanium diode. Similarly, Fig. 3-1(b) displays the forward *I–V* characteristics for a silicon diode. Note that forward current I_F increases slowly at first until forward-bias voltage V_F overcomes the junction potential, after which current increases rapidly. The value of voltage to overcome the junction potential depends on the semiconductor material. This value is approximately 0.15 to 0.3 V for germanium diodes and 0.5 to 0.7 V for silicon diodes.

3-1.2 Reverse Characteristics. When a reverse-bias voltage is applied across a diode, it conducts a small saturation current, I_s, until the diode breaks down. Figure 3-2(a) shows typical reverse characteristics of a germanium diode; Fig. 3-2(b) shows reverse characteristics for a silicon diode. The saturation current for a germanium diode is of the order of microamperes (10^{-6} A); that for a silicon diode, of nanoamperes (10^{-9} A). In Section 2-2.2

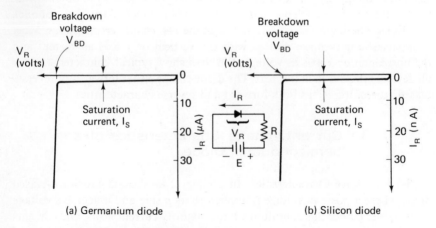

Figure 3-2 Typical reverse characteristics of semiconductor diodes.

we discussed how saturation current is independent of reverse-bias voltage for values less than the breakdown voltage. Figure 3-2 clearly illustrates this property. However, saturation current does depend on temperature; that is, as temperature increases, saturation current increases. Since the saturation current of a silicon diode is approximately 1000 times smaller than that of a germanium diode, we should use silicon diodes wherever possible.

We see from the characteristics that after breakdown voltage is reached, current increases very rapidly. This current increase must be limited by an external resistor or the diode will be destroyed. The value of breakdown voltage V_{BD} in Fig. 3-2 is not specified. This is because almost any value of breakdown voltage greater than 3 V may be obtained, depending on how the diode is manufactured.

3-1.3 Operating Regions. Figure 3-3 is a combination of Figs. 3-1 and 3-2. It is a plot of both forward and reverse *I–V* characteristics. Unlike Fig. 3-2, the scale for reverse current I_R is plotted in milliamperes, and for this reason I_s approximates the horizontal axis.

Figure 3-3 shows two operating regions:

1. The *rectifying region*, which extends along the *I–V* curve from the breakdown voltage into the forward-bias region.

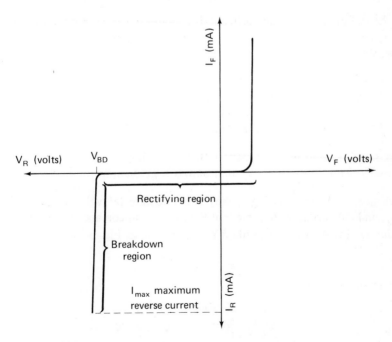

Figure 3-3 Forward and reverse characteristics of a diode.

2. The *breakdown region*, which lies between the saturation current at V_{BD} and I_{max}.

The calculation of I_{max} will be discussed in Section 3-4.3. Diodes that are operated in the rectifying region are called *rectifying diodes*; those that are operated in the breakdown region are referred to as *Zener diodes.*

The word "rectify" means to pass current in only one direction. When a diode is forward-biased, current I_F flows easily through the diode, producing a fixed voltage drop across it from anode to cathode. For a germanium diode this voltage drop is approximately 0.3 V; for a silicon diode, 0.6 V. When a diode is reverse-biased, the current, for all practical purposes, is zero. Zero current means no current flow, which is the same as an open circuit. Therefore, in any circuit that contains a reverse-biased diode, think of the diode as an open connection to the remaining circuit. These ideas for forward- and reverse-biased diodes are extremely helpful in analyzing circuits that contain diodes and will be used in the next sections.

The diode symbol introduced in Section 2-4 and Fig. 3-1 is the symbol for rectifying diodes. The Zener diode symbol along with the properties and applications of Zener diodes will be studied in Section 3-4.

3-2 Diode Circuits with DC Sources

Most applications that use rectifying diodes have ac input voltages. However, let us reconsider applying a dc source to a diode circuit to emphasize the following points:

1. Kirchhoff's voltage law.
2. Diode voltage drop.
3. Voltage drop across the load resistor.
4. Value of current in the circuit.

For simplification we may assume that the voltage drop across a forward-biased silicon diode is 0.6 V; that across a germanium diode, 0.2 V.

Example 3-1: Refer to Fig. 3-4(a) and calculate (a) voltage across R_L or V_{RL}, and (b) current I. Assume that the diode is silicon and that $V_D = 0.6$ V. *Solution:* (a) Applying Kirchhoff's voltage law to Fig. 3-4(a),

$$E = V_D + V_{RL} \tag{3-1}$$

Rearranging terms,

$$V_{RL} = E - V_D$$

and

$$V_{RL} = 10 \text{ V} - 0.6 \text{ V} = 9.4 \text{ V}$$

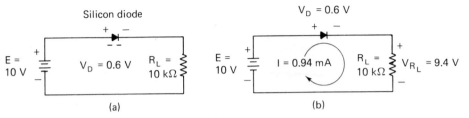

Figure 3-4 Circuit diagrams for Example 3-1.

(b) From Ohm's law,

$$I = \frac{V_{RL}}{R_L} = \frac{9.4 \text{ V}}{10 \text{ k}\Omega} = 0.94 \text{ mA}$$

Thus most of the input voltage is across R_L and R_L controls or limits the current, as shown in Fig. 3-4(b).

Example 3-2: Voltage source E in Fig. 3-5(a) reverse-biases the diode. Calculate (a) voltage across R_L, V_{RL}, and (b) voltage across the diode V_D. Assume a saturation current of 10 nA (10×10^{-9} A).
Solution:

(a) $V_{RL} = I_s R_L = (10 \times 10^{-9} \text{ A})(10 \times 10^3 \ \Omega) = 100 \times 10^{-6} \text{ V}$
  $= 100 \ \mu\text{V}$

(b) Rearranging Eq. (3-1),

$$V_D = E - V_{RL} = 10 \text{ V} - 100 \times 10^{-6} \text{ V} \approx 10 \text{ V}$$

The voltage drop across R_L is so small that essentially all of E appears across the diode. The polarity of the voltages are given in Fig. 3-5(b). In this type of circuit the breakdown voltage V_{BD} of the rectifying diode should always be greater than the input voltage E.

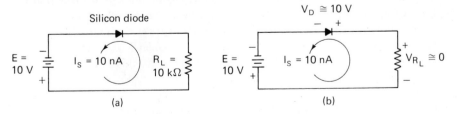

Figure 3-5 Circuit diagrams for Example 3-2.

3-3 Diode Circuits with AC Sources

3-3.1 Half-Wave Rectifier Circuit. One of the fundamental applications of a diode is the half-wave rectifier circuit of Fig. 3-6. This circuit is used in some dc power supplies. The input voltage is an ac sine wave of voltage. In the dc circuits studied in Section 3-2, the diode was always forward- or reverse-biased, depending on the polarity of the source. With an ac source the polarity is continually changing. This results in the diode being forward-biased, and then as the polarity of the source changes, the diode becomes reverse-biased. During the positive half-cycles, the diode is forward-biased, which results in a voltage drop across the silicon diode of approximately 0.6 V. The peak output voltage during these positive cycles is 10 V − 0.6 V = 9.4 V. This is similar to the dc diode circuit of Fig. 3-4(a). During the negative half-cycles of input voltage, the diode is reverse-biased and only the saturation current can flow. For a silicon diode the saturation current is in the order of 10 nA (10×10^{-9} A). Therefore, the voltage drop across R_L is (10 kΩ)(10 nA) = 100 μV ≈ 0. The output voltage during the negative half-cycles is so small that it can be considered zero. This situation is similar to the reverse bias of the dc circuit of Fig. 3-5(b). Figure 3-6(c) is the output-voltage waveform of a half-wave rectifier circuit.

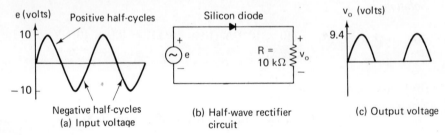

Figure 3-6   Half-wave rectification.

3-3.2 Full-Wave Rectifier Circuit. Although some dc power supplies use the half-wave rectifier, the full-wave rectifier circuit of Fig. 3-7(b) is the basic building block of most dc supplies. Our analysis of this circuit, like that of the half-wave rectifier, is to obtain the output-voltage waveform.

For positive cycles diodes 1 and 2 are forward-biased, diodes 3 and 4 are reverse-biased. To help in the understanding, consider reverse-biased diodes as an open circuit. Figure 3-7(e) is drawn to illustrate this point for positive cycles. The path of current during positive cycles is drawn on Fig. 3-7(e). Note the direction of the diode triangle points in the direction of conventional current flow. Diodes 1 and 2 are forward-biased, each having a voltage drop of approximately 0.6 V. Therefore, the peak output voltage is 10 V − 0.6 V − 0.6 V = 8.8 V.

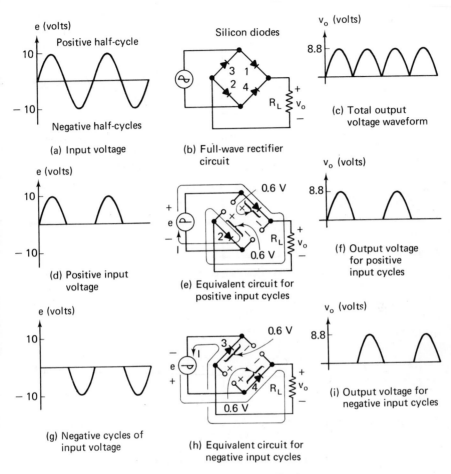

Figure 3-7 Full-wave rectification.

During negative cycles of the input voltage diodes 1 and 2 are reverse-biased, diodes 3 and 4 are forward-biased. Figure 3-7(h) is drawn to show the path of current for negative cycles. Diodes 1 and 2 are not drawn because they are reverse-biased. The direction of current through R_L is the same as that of Fig. 3-7(e). Therefore, although the input voltage is negative, the output voltage is positive. Once again the peak output voltage is 10 V − 0.6 V − 0.6 V = 8.8 V. The total output voltage is shown in Fig. 3-7(c).

3-3.3 Clipping Circuits. Clipping circuits receive an input voltage and clip (or cut off) the positive and/or negative cycles. Figure 3-8(b) is a clipping circuit; the output is taken across two silicon diodes. For positive half-cycles diode 1 is forward-biased and diode 2 is reverse-biased. Figure 3-8(e) is drawn to illustrate the idea that a reverse-biased diode may be considered

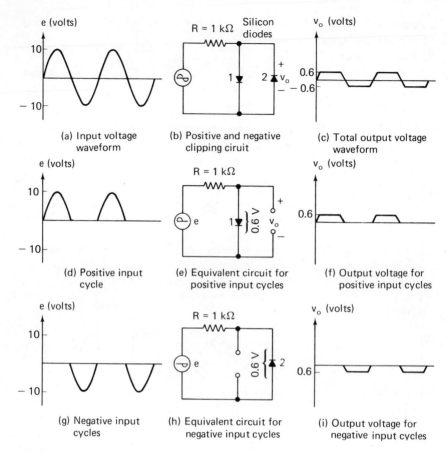

Figure 3-8 Examples of clipping circuits.

open. When the input voltage reaches 0.6 V, diode 1 is forward-biased, and the output that is taken across diode 1 remains at 0.6 V. Thus the input voltage is clipped to produce the output voltage [see Fig. 3-8(f)]. For negative half-cycles, diode 1 is reverse-biased and diode 2 is forward-biased. The simplified circuit of Fig. 3-8(h) shows that the negative input cycles will be clipped by diode 2. Since diode 2 is a silicon diode, the output voltage will remain at 0.6 V, as shown in Fig. 3-8(i). Figure 3-8(c) is the combination of Fig. 3-8(f) and (i) to yield the total output-voltage wave form.

3-4 Zener Diodes

3-4.1 Circuit Symbol. The two methods that cause junction breakdown were discussed in Section 2-3. These methods are avalanche breakdown and junction breakdown. Regardless of the type of breakdown, the common name applied to any semiconductor diode operating in the breakdown region is *Zener diode*. Other names, which are not used very often, are "avalanche diode" and "breakdown diode."

As shown in Fig. 3-9, the Zener diode's symbol differs from that of the rectifying diode. The Zener is designed to be operated in its breakdown region and, as will be shown, it normally conducts current in the *reverse* direction.

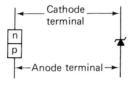

(a) pn junction   (b) Zener diode
circuit symbol

Figure 3-9   Zener diode symbol.

3-4.2 Current–Voltage Characteristics of a Zener Diode. Figure 3-10 shows the forward and reverse characteristics of a Zener diode. Note that

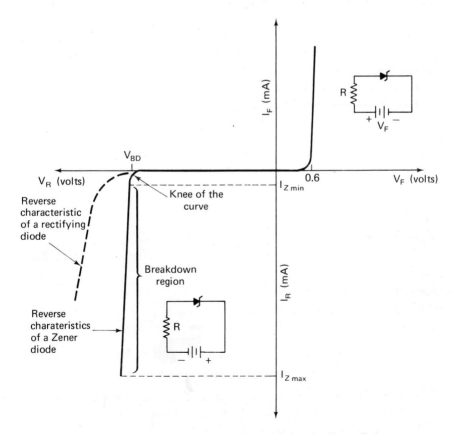

Figure 3-10   Forward and reverse characteristics of a Zener diode.

the forward characteristics are identical with that of any silicon rectifying diode. Specifically, it takes approximately 0.6 V to overcome the junction potential, after which I_F increases rapidly and must be limited by the external circuit. In the reverse direction up to breakdown voltage V_{BD}, only negligible saturation current flows. This value of current is of the order of 10 nA for both silicon-rectifying diodes and Zener diodes. Figure 3-10 compares the rapid breakdown of a Zener diode with that of a rectifying diode. The dotted line is a typical reverse characteristic of a rectifying diode, and the solid line is for a Zener diode. Once reverse-bias voltage across a Zener diode reaches V_{BD}, current increases so rapidly that we must be able to control it by the external circuit; otherwise, the diode will be destroyed. To be able to operate a Zener diode in the breakdown region without destroying it, we must know what maximum current the Zener can handle.

3-4.3 Maximum Zener Current. A principal cause of the destruction of electronic components is heat. That is, when we apply power to a device, the device heats up. If we apply too much power, the device is destroyed. Therefore, device manufacturers specify maximum power capability, P_{Dmax}, of a device and give this information on their data sheets. Zener diodes may be purchased from $\frac{1}{4}$ W (250 mW) to 50 W. Besides P_{Dmax}, Zener-diode manufacturers give the Zener's voltage, V_Z, which for all practical applications is the breakdown voltage, V_{BD}. Since power is the product of voltage times current,

$$P_{Dmax} = V_Z I_{Zmax} \tag{3-2}$$

where I_{Zmax} is the maximum reverse current that can flow without destroying the Zener. Rearranging Eq. (3-2) yields

$$I_{Zmax} = \frac{P_{Dmax}}{V_Z} \tag{3-3}$$

Example 3-3: A 1N5221 Zener diode has the following specifications: $V_Z = 2.4$ V and $P_{Dmax} = 500$ mW. Calculate the maximum allowable Zener current.
Solution: Applying Eq. (3-3),

$$I_{Zmax} = \frac{500 \text{ mW}}{2.4 \text{ } V} \approx 208 \text{ mA}$$

Example 3-4: Calculate the maximum Zener current for a 1N5281 Zener diode; $V_Z = 200$ V and $P_{Dmax} = 500$ mW.
Solution: Using Eq. (3-3),

$$I_{Zmax} = \frac{500 \text{ mW}}{200 \text{ V}} = 2.5 \text{ mA}$$

Both the 1N5221 and the 1N5281 are from the same series. This means that both diodes have the same value of P_{Dmax} and the same outside case but not the same Zener voltage. Clearly one cannot substitute one Zener diode for another without knowing both V_Z and P_{Dmax}.

3-4.4 Zener Diode Under Operating Conditions. When operating a Zener diode in the breakdown region we should be aware that:

1. The external voltage used to reverse bias the Zener must be greater than the breakdown voltage. When this occurs, the voltage across the Zener is constant at V_Z.
2. The current through the Zener must be less than I_{Zmax}.
3. There is a minimum value of current that must flow through the Zener to keep it in the breakdown region. For low-voltage Zeners this value is about 5 mA.

Example 3-5: For the Zener diode of Fig. 3-11, find (a) I_{Zmax} and (b) current through the Zener when connected into the circuit.

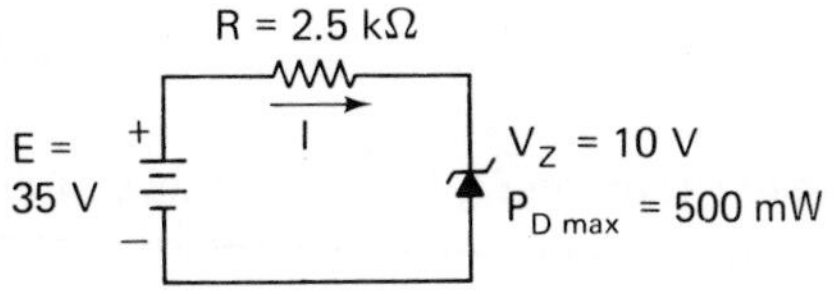

$$R = 2.5 \text{ k}\Omega$$

Figure 3-11 Circuit diagram for Example 3-5.

$E = 35 \text{ V}$ I $V_Z = 10 \text{ V}$ $P_{D \max} = 500 \text{ mW}$

Solution:

(a) Applying Eq. (3-3),

$$I_{Zmax} = \frac{500 \text{ mW}}{10 \text{ V}} = 50 \text{ mA}$$

(b) Since the applied voltage E is greater than V_Z, the Zener is operating in the breakdown region and the voltage across the Zener is V_Z. The voltage across the 5-kΩ resistor is $E - V_Z = 35 \text{ V} - 10 \text{ V} = 25 \text{ V}$, and the current through the 2.5 kΩ is

$$I = \frac{25 \text{ V}}{2.5 \text{ k}\Omega} = 10 \text{ mA}$$

Current $I = 10 \text{ mA}$ is also the current through the Zener because the circuit of Fig. 3-11 is a series circuit. Note that the operating current 10 mA is less than I_{Zmax} and its value is controlled by R. For example, if R is changed to 1 kΩ, the current now flowing through the Zener is

$$I = \frac{25 \text{ V}}{1 \text{ k}\Omega} = 25 \text{ mA}$$

Example 3-6: Refer to Fig. 3-11 and determine the minimum value of R to keep the Zener from burning out.

Solution: Remember that to keep the Zener from being destroyed, the current through the Zener has to be less than I_{Zmax}. From Example 3-5(a), I_{Zmax} = 50 mA; therefore, the minimum value of R is

$$R_{min} = \frac{35\ V - 10\ V}{50\ mA} = 0.5\ k\Omega = 500\ \Omega$$

3-4.5 Minimum Zener Current. Approximately 5 mA is needed to keep many types of Zener diodes operating below the knee of the curve in the breakdown region. This minimum value of current as well as the maximum value of current in the circuit of Fig. 3-11 is controlled by resistor R.

Example 3-7: Calculate the value of R in Fig. 3-11 to guarantee that at least 5 mA flows through the Zener.

Solution: Since $V_Z = 10\ V$,

$$R_{max} = \frac{35\ V - 10\ V}{5\ mA} = \frac{25\ V}{5mA} = 5\ k\Omega$$

If R is greater than 5 kΩ, less than 5 mA of current flows through the Zener and results in the Zener no longer operating reliably in the breakdown region.

3-4.6 Voltage Regulation with Load Variation. A principal application of a Zener diode is *voltage regulation*. In this type of circuit a load resistor is connected across the Zener. With the Zener diode operating in the breakdown region, it is capable of maintaining a constant voltage across the load. Thus the Zener acts as an inexpensive battery. The value of voltage across the load resistor is the Zener voltage, V_Z. (*Note:* To keep the Zener in the breakdown region, the current through the Zener must be between I_{Zmin} and I_{Zmax}.)

Example 3-8: For the circuit of Fig. 3-12, calculate (a) total current I, (b) load current I_L, and (c) Zener current I_Z.

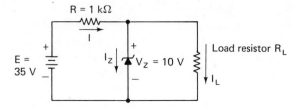

Figure 3-12 Voltage-regulator circuit of Example 3-8.

Solution:

(a) $I = \dfrac{E - V_Z}{R} = \dfrac{35 \text{ V} - 10 \text{ V}}{1 \text{ k}\Omega} = 25 \text{ mA}$

(b) The Zener and load resistor are in parallel; therefore, the voltage across R_L is V_Z, and then

$$I_L = \frac{V_Z}{R_L} = \frac{10 \text{ V}}{500 \text{ }\Omega} = 20 \text{ mA}$$

(c) Applying Kirchhoff's current law,

$$I_Z = I - I_L = 25 \text{ mA} - 20 \text{ mA} = 5 \text{ mA}$$

Unlike Examples 3-5 and 3-6, the total current I is not the Zener current. In the circuit of Fig. 3-12 the Zener current is the difference between the total current I and the load current I_L.

Example 3-9: Refer to Fig. 3-12 and let $R_L = 5 \text{ k}\Omega$. Again calculate (a) total current I, (b) load current I_L, and (c) Zener current I_Z.
Solution:

(a) $I = \dfrac{E - V_Z}{R} = \dfrac{35 \text{ V} - 10 \text{ V}}{1 \text{ k}\Omega} = 25 \text{ mA}$

(b) $I_L = \dfrac{V_Z}{R_L} = \dfrac{10 \text{ V}}{5 \text{ k}\Omega} = 2 \text{ mA}$

(c) $I_Z = I - I_L = 25 \text{ mA} - 2 \text{ mA} = 23 \text{ mA}$

Now, comparing answers of Example 3-8 with those of Example 3-9, we see that the total current remains unchanged. The voltage across the load also remains constant at the Zener voltage. The current through the Zener is $I - I_L$; therefore, the current through the Zener is the amount of current that the load does not need. In conclusion, the load resistor changed from 500 Ω to 5 kΩ, but the voltage across it remained constant (or was regulated) at 10 V. The Zener absorbed any change in load current to maintain a constant output voltage.

3-4.7 Voltage Regulation with Input Variation. In this application, one value of load resistor is connected across the Zener and input voltage E is varied. As long as E is greater than V_Z and Zener current is between $I_{Z\text{min}}$ and $I_{Z\text{max}}$, the Zener will be operating in the breakdown region and load voltage will remain constant.

Example 3-10: For the circuit of Fig. 3-13, calculate (a) total current I, (b) load current I_L, and (c) Zener current.

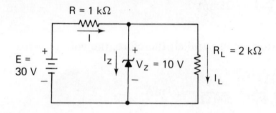

Figure 3-13 Voltage-regulator circuit of Example 3-10.

Solution:

(a) $I = \dfrac{E - V_z}{R} = \dfrac{30\text{ V} - 10\text{ V}}{1\text{ k}\Omega} = 20\text{ mA}$

(b) $I_L = \dfrac{V_z}{R_L} = \dfrac{10\text{ V}}{2\text{ k}\Omega} = 5\text{ mA}$

(c) $I_z = I - I_L = 20\text{ mA} - 5\text{ mA} = 15\text{ mA}$

Example 3-11: Refer to Fig. 3-13 and let $E = 40$ V. Calculate (a) I, (b) I_L, and (c) I_Z.
Solution:

(a) $I = \dfrac{E - V_z}{R} = \dfrac{40\text{ V} - 10\text{ V}}{1\text{ k}\Omega} = 30\text{ mA}$

(b) $I_L = \dfrac{V_z}{R_L} = \dfrac{10\text{ V}}{2\text{ k}\Omega} = 5\text{ mA}$

(c) $I_z = I - I_L = 30\text{ mA} - 5\text{ mA} = 25\text{ mA}$

In comparing answers of Examples 3-10 and 3-11 we see that because E changed, the total current I had to change. The load current I_L did not change because both the Zener voltage and load resistor remained constant. Once again the current through the Zener is that amount of current which the load did not need.

3-4.8 Zener Diode Resistance. Up to this point we have indicated that once the voltage across the Zener reaches the breakdown voltage, there is no further increase in Zener voltage. This is not precisely correct, although we shall now show that this approximation is valid for most practical applications. What does happen is that the Zener's voltage increases slightly as current through the Zener increases. This increase in voltage is noticeable on the characteristics because at the breakdown voltage the reverse characteristic is not vertical. Figure 3-14 shows that the curve has a slope from V_{BD}

to I_{Zmax}. This slope is the Zener's resistance and be found graphically from

$$r_z = \frac{\Delta V_z}{\Delta I_z} \frac{\text{change in Zener voltage}}{\text{change in Zener current}} \qquad (3\text{-}4)$$

For example, in Fig. 3-14, since $\Delta V_z = 0.5$ V and $\Delta I_z = 20$ mA,

$$r_z = \frac{\Delta V_z}{\Delta I_z} = \frac{0.5\ V}{20\ \text{mA}} = 25\ \Omega$$

Typical values of r_z are from 2 to 40 Ω.

Figure 3-15 shows that the Zener diode may be replaced by an equivalent circuit of a battery (whose value is V_{BD}) in series with the Zener resistance (whose value is r_z). The Zener voltage V_Z is now V_{BD} plus the voltage drop across r_z, or, in equation form,

$$V_Z = V_{BD} + I_Z r_z \qquad (3\text{-}5)$$

where V_{BD} is the breakdown voltage and I_Z is the current through the Zener for a particular application.

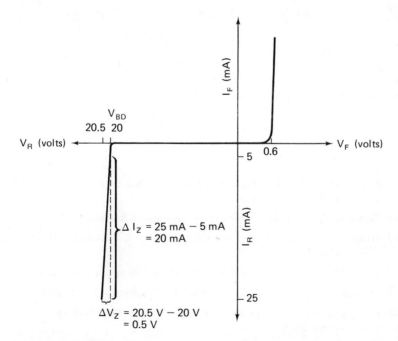

Figure 3-14 Shape of reverse characteristics in the breakdown region is the Zener resistance.

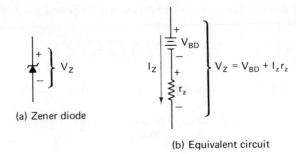

(a) Zener diode

(b) Equivalent circuit

Figure 3-15 Zener diode model.

Example 3-12: If the breakdown voltage is 20 V and r_z has been calculated from Eq. (3-4) to be 25 Ω, what is the Zener voltage with 10 mA flowing through it?

Solution: From Eq. (3-5),

$$V_Z = 20 \text{ V} + (10 \text{ mA})(25 \text{ Ω}) = 20 \text{ V} + 0.25 \text{ V} = 20.25 \text{ V}$$

If we increase I_Z to 20 mA, the Zener voltage is

$$V_Z = 20 \text{ V} + (20 \text{ mA})(25 \text{ Ω}) = 20 \text{ V} + 0.5 \text{ V} = 20.5 \text{ V}$$

Thus the voltage drop $I_Z r_z$ is small, and for most practical applications $V_Z \approx V_{BD}$ is satisfactory. Note that the smaller the value of r_z, the smaller the value of $I_Z r_z$. This means that if a Zener is being used as a voltage regulator, try to purchase a Zener with small r_z. Then the load voltage will not change as the Zener current changes. This principle will be utilized in Chapter 17.

Problems

3-1 Approximately what voltage drop exists across a forward-biased (a) silicon diode; (b) germanium diode?

3-2 Sketch a typical silicon diode current–voltage characteristic.

3-3 When a rectifying diode is reverse-biased, does it behave like an open or a closed switch?

3-4 Does a forward-biased rectifying diode behave like an open or a closed switch?

3-5 Voltage E is increased to 20 V in Example 3-1. Find (a) V_{RL}; (b) I.

3-6 E is increased to 20 V in Example 3-2. Does (a) diode voltage double; (b) diode current double?

3-7 Ac voltage e has a peak value of 15.6 V in Fig. 3-6(a). What is the peak output voltage?

3-8 In Fig. 3-7, *e* has peak values of 15.6 V. What is the peak output voltage?

3-9 Which diodes conduct in Fig. 3-7(a) during the negative-input half-cycle?

3-10 Do diodes D3 and D4 of Fig. 3-7(a) act as open or closed switches during the positive-input half-cycle?

3-11 What type of diode operates on the principle of avalanche or junction breakdown?

3-12 Show the output-voltage wave form in Fig. 3-8 if diode 2 is an open circuit.

3-13 Show the output-voltage wave form in Fig. 3-8 if diode 2 is short-circuited.

3-14 A 5-V Zener is rated at 500 mW. Calculate maximum Zener current.

3-15 Find the Zener current in Fig. 3-13 when $E = 15$ V.

3-16 What is the Zener power in Problem 3-15?

3-17 In Example 3-6, find minimum R if $E = 15$ V.

3-18 If $E = 15$ V in Fig. 3-11, find the value of R to guarantee 5 mA through the Zener.

3-19 In Fig. 3-12, E is reduced to 25 V and $R_L = 1$ kΩ. Find (a) total current I; (b) I_L; (c) I_Z.

3-20 In Fig. 3-13, $E = 20$ V. Find (a) I; (b) I_L; (c) I_Z. Compare the results with Examples 3-10 and 3-11.

4

Introduction to the Bipolar
Junction Transistor

4-0 Introduction

The bipolar junction transistor, BJT, is one of the most widely used semiconductor devices. Its name tells quite a bit about it. Bipolar signifies that there are two types of BJT. One depends primarily on holes for conduction, *pnp*, and the other on electrons, *npn*. So bipolar means two polarities of current carriers. The term "junction" refers to the two *pn* junctions present in all BJTSs. Finally, the word "transistor" describes one job that the BJT does best. That is, the BJT *transforms resistance*.

While the idea of transforming resistance may seem at first to be bewildering, it will turn out eventually to be quite easy and will in fact simplify enormously the task of circuit analysis. But before resistance transformation can be studied, we must build a foundation beginning with transistor construction.

4-1 BJT Structure and Circuit Symbols

4-1.1 Structure of a BJT. The construction essentials of one type of BJT are shown in Fig. 4-1(a). The structure is not drawn to scale because the base region is actually very thin, in the order of one millionth of an inch. The BJT consists of three layers of alternately doped semiconductor material labeled *emitter*, *base*, and *collector*. The emitter *emits* or furnishes a supply of charge carriers that can be *collected* by the collector and *controlled* by the base. Leads are bonded or soldered to the emitter, base, and collector by

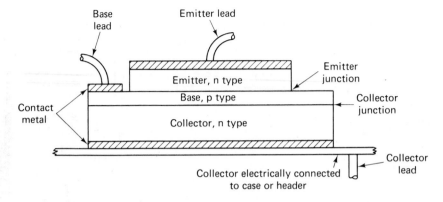

(a) Transistor structure

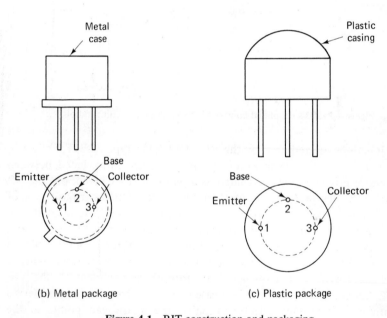

(b) Metal package   (c) Plastic package

Figure 4-1 BJT construction and packaging.

means of contact metal pads. The structure is encased in a package. One type of metal package is shown in Fig. 4-1(b) and one type of plastic package in Fig. 4-1(c).

4-1.2 BJT Circuit Symbols. To explain transistor action it is customary and easier to sketch the BJT as in Fig. 4-2(a) or (b) rather than as in Fig. 4-1(a). When representing the BJT in a schematic drawing, use the circuit symbols of Fig. 4-2(c) or (d).

Both Figs. 4-1(a) and 4-2(a) and (b) indicate that three lead terminals and

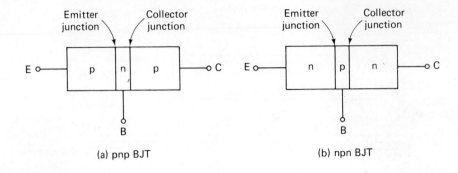

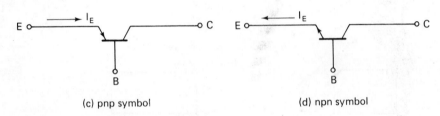

Figure 4-2 BJT simplified construction sketch and circuit symbols.

two *pn* junctions are required by the BJT. One basic type of BJT, shown in Fig. 4-2(a), has one layer of *n*-doped semiconductor sandwiched between two layers of *p*-type and is designated as a *pnp* BJT. By convention, the left or first *p* in the designation *pnp* refers to the emitter. The other basic type of BJT in Fig. 4-2(b) has a *p* layer between two *n* layers and is designated *npn*.

Circuit symbols for *pnp* and *npn* transistors are given in Fig. 4-2(c) and (d), respectively. The arrowhead in the symbol gives three important bits of information. First, the arrowhead points toward the *n*-type material—*n*-type base for *pnp*, *n*-type emitter for *npn*. Therefore, the arrow identifies the type of BJT. Second, the arrowhead is located on and identifies the emitter to differentiate it from the collector. Finally, the arrowhead points in the direction that conventional current will flow. Conventional current flow will be used throughout this text to agree with the arrowhead direction.

4-2 Operating Methods for the BJT

It is evident from Figs. 4-1 and 4-2 that both *pnp* and *npn* transistors have two *pn* junctions, an emitter junction and a collector junction. There are two possible ways of operating a *pn* junction. It can either be forward-biased or reverse-biased. Since there are two junctions and each can be independently forward- or reverse-biased, there are four methods of operating the BJT. These methods of operating the BJT are often called *operating modes,* and each has a name as shown in Table 4-1.

Table **4-1** BJT OPERATING MODES

Emitter-Junction Bias	Collector-Junction Bias	Operating Mode
Forward	Forward	Saturation
Forward	Reverse	Linear or active
Reverse	Forward	Inverse
Reverse	Reverse	Cutoff

BJTs are rarely operated in the inverse mode. As will be shown, cutoff and saturated modes of operating make the BJT behave like a switch, that is, either off or on. We shall begin our study of transistor action with the linear or active mode.

In the linear mode the BJT performs as an amplifier. That is, a weak input signal is applied between two of its three terminals and a stronger or amplified version of the input signal is developed between the third terminal and one of the input terminals. One terminal is common to both input and output. This point is introduced now but will be dealt with squarely in Section 4-8.

It is helpful to review operating modes by drawing an analogy between the BJT and a water control valve. BJT cutoff corresponds to a closed valve. BJT linear mode corresponds to the valve being partially open. A fully open valve is analogous to BJT saturation. That is the valve allows maximum water flow and the BJT allows maximum current flow.

4-3 *pnp* Transistor Action

4-3.1 Internal Transistor Currents. Basic transistor action is shown in Fig. 4-3. Battery V_{BE} provides forward bias for the emitter junction. Since the emitter is made of p-type material, holes flow from the emitter across the forward-biased emitter junction. The base region is very thin, so thin in fact that roughly 99% of these holes diffuse across the base region and enter the space-charge region that surrounds the collector junction. Holes are then accelerated across the collector junction's space-charge region because of its reverse bias from battery V_{CC}. Recall from Section 2-1.3 that holes in the n side (base) of a reverse-biased junction are minority carriers and are swept across the junction to the p side (collector) by the electric field.

Since resistance of base and emitter semiconductor material is small, just about all of V_{BE} is developed across the emitter junction. The collector

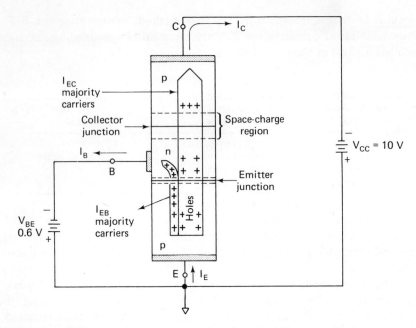

Figure 4-3 Internal *pnp* transistor currents (silicon transistor).

resistance is so small that the collector junction is reverse-biased by $V_{CC} - V_{BE} = 10 - 0.6 = 9.4$ V.

Internal transistor current I_{EC} consists of holes flowing from emitter to collector. It is the job of emitter and collector ohmic contacts to exchange electrons for holes so that only electrons will flow in the external circuit (see Section 1-8). That is, electrons will be furnished by the negative terminal of V_{CC} to the collector terminal, and electrons will be returned from the emitter terminal to the positive terminal of V_{CC}.

Internal current I_{EB} consists of holes fed from the emitter that do not reach the collector but become neutralized by combining with electrons fed into the base, by the base ohmic contact. These electrons are furnished by the negative terminal of V_{BE}.

4-3.2 External Transistor Currents. We cannot easily go inside the transistor to measure its internal currents. We can, however, measure currents at the BJT's terminals. To focus attention on these external (to the transistor) circuit currents, the BJT model of Fig. 4-3 is replaced by the BJT circuit symbol in Fig. 4-4. Here we use conventional current (the opposite of electron current) because the emitter arrow points in the direction of conventional current flow when the emitter junction is forward-biased. Current enters the transistor via the emitter terminal. Current leaves the transistor via both base and collector terminals. Since current I_E entering the transistor must

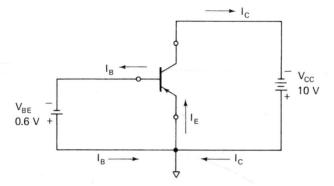

Figure 4-4 External *pnp* transistor circuit currents.

equal total current $I_B + I_C$ leaving the transistor, we can write

$$I_E = I_B + I_C \tag{4-1}$$

where I_E = dc emitter current
I_B = dc base current
I_C = dc collector current

4-3.3 Transistor Beta. In Fig. 4-3 external dc base current I_B equals internal transistor current I_{EB}. External dc collector current I_C equals internal current I_{EC}. The ratio of collector current I_C to base current I_B is an extremely important electrical characteristic of the BJT. There are two symbols for this ratio, β or h_{FE}. Manufacturing data sheets specify the ratio I_C/I_B as h_{FE}. However, the earlier term β is easier to write or speak and is commonly found in magazine articles and textbooks. Accordingly, this text will use the symbol β and define it as

$$\left(\beta = \frac{I_C}{I_B} \right) \tag{4-2}$$

A typical value for β is 50, and it may range from a low of 10 to a high value that can exceed 600. Measurement of β and how it is used will be dealt with in Section 5-3.

4-3.4 Transistor Alpha. Another important transistor characteristic is the ratio of holes reaching the collector to holes leaving the emitter. From Fig. 4-3 this ratio is seen to be the ratio of I_{EC} to $I_{EC} + I_{EB}$. Since $I_{EC} = I_C$ and $I_{EC} + I_{EB} = I_E$, it is easier to express the ratio as I_C/I_E. There are two symbols for this ratio. On manufacturers' data sheets for BJTs the ratio is symbolized by h_{FB}. We shall use the earlier term α and define it as

$$\alpha = \frac{I_C}{I_E} \tag{4-3}$$

From Fig. 4-3, I_{EB} or I_B is defined in terms of I_E by

$$(1 - \alpha)I_E = I_B \text{ ?}. \tag{4-4}$$

Typical values of α range from 0.980 to 0.998. Since very few holes fail to cross the base region in Fig. 4-3, we would expect that I_C will approximately equal I_E and that α will always very nearly equal 1. It is difficult to measure α with any precision, because it is the ratio of two very nearly equal quantities. It is easy to measure β, however, and values for β rather than α are given on manufacturers' data sheets. In the next section we shall see how α and β are related so that α can be calculated if values for β are known.

4-4 Relations Between Transistor Terminal Currents

Internal transistor currents in Fig. 4-3 are extended as loop currents to the BJT terminals in Fig. 4-5. I_C and I_B may be expressed in terms of α and I_E. Their relationships are shown in Fig. 4-5(b) and also by Eqs. (4-1) and (4-2), repeated here for convenience.

$$(4\text{-}1) \quad I_E = I_B + I_C \qquad (4\text{-}2) \quad I_C = \beta I_B$$

Note that I_C in Eq. (4-2) is expressed in terms of I_B. By substituting for I_C from Eq. (4-2) into (4-1) we express I_E in terms of I_B as

$$I_E = I_B + \beta I_B$$

or
$$I_E = (\beta + 1)I_B \tag{4-5}$$

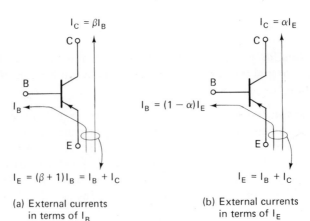

(a) External currents in terms of I_B

(b) External currents in terms of I_E

Figure 4-5 External transistor currents.

4-5 Relations Between Alpha and Beta

To see how α is related to β, obtain the ratio of Eq. (4-2) to Eq. (4-5):

$$\frac{I_C}{I_E} = \frac{\beta I_B}{(\beta + 1)I_B} \quad \text{or} \quad \frac{I_C}{I_E} = \frac{\beta}{\beta + 1}$$

So the ratio I_C/I_E can be expressed in two ways:

$$\frac{I_C}{I_E} = \alpha = \frac{\beta}{\beta + 1}$$

Thus

$$\alpha = \frac{\beta}{\beta + 1} \qquad (4\text{-}6)$$

Another useful relationship between α and β is found by obtaining the ratio of I_C to I_B from Fig. 4-5(b):

$$\frac{I_C}{I_B} = \frac{\alpha I_E}{(1 - \alpha)I_E} = \frac{\alpha}{1 - \alpha}$$

The ratio of I_C to I_B from Fig. 4-5(a) is

$$\frac{I_C}{I_B} = \frac{\beta I_B}{I_B} = \beta$$

Therefore,

$$\beta = \frac{\alpha}{1 - \alpha} \qquad (4\text{-}7)$$

Equation (4-7) expresses β in terms of α.

Example 4-1: If $I_B = 20 \ \mu A$ and β is given as 49, find I_C and I_E.
Solution: From Eq. (4-2), $I_C = \beta I_B = 49 \times 20 \ \mu A = 980 \ \mu A$. From Eq. (4-5), $I_E = (\beta + 1)I_B = 50 \times 20 \ \mu A = 1000 \ \mu A = 1 \ mA$; or, from Eq. (4-1), $I_E = I_B + I_C = (20 + 980) \ \mu A = 1000 \ \mu A = 1 \ mA$.

Example 4-2: If $\beta = 49$ for a BJT, find α.
Solution: From Eq. (4-6),

$$\alpha = \frac{49}{49 + 1} = 0.980$$

Example 4-3: If $\alpha = 0.980$, calculate β and compare the results with Example 4-2.

Solution: From Eq. (4-7),

$$\beta = \frac{0.980}{1.000 - 0.980} = \frac{0.980}{0.020} = 49$$

The above examples show how to find I_C and I_E if I_B and either α or β are known. Two more examples are selected to show how to find I_B and I_C if I_E and either α and β are known.

Example 4-4: If $I_E = 1$ mA $= 1000$ μA and $\beta = 49$, find I_B and I_C.
Solution: Rearranging Eq. (4-5),

$$I_B = \frac{I_E}{\beta + 1} = \frac{1000\ \mu A}{50} = 20\mu A$$

From Eq. (4-2),

$$I_C = 49 \times 20\ \mu A = 980\ \mu A$$

Example 4-5: If $I_E = 1$ mA $= 1000$ μA and $\alpha = 0.980$, find I_B and I_C.
Solution: From Eq. (4-3),

$$I_C = 0.980 \times 1000\ \mu A = 980\ \mu A$$

From Eq. (4-1), $I_B = I_E - I_C = 1000\ \mu A = 980\ \mu A = 20\ \mu A$.

Compare Examples 4-1 through 4-5 to see how they all are merely different ways of expressing the same current relationships in the same transistor. The conclusion to be drawn is that if we can evaluate one BJT terminal current in a circuit and know the internal BJT current split (by either α or β), we can immediately determine the other BJT currents.

4-6 *npn* Transistor Action

Compare *pnp* transistor action in Figs. 4-3 to 4-5 with *npn* transistor action in Fig. 4-6 to see that (1) currents flow in opposite directions, (2) external supply voltages are reversed from one another, but (3) the relationships between transistor currents are identical.

In Fig. 4-6(a), V_{BE} forward-biases the emitter junction to allow the *n*-type emitter to inject majority electrons into the base. Most of these electrons drift across the thin base region as I_{EC} and are accelerated across the reverse-biased collector junction into the collector body. Here they are collected by the positive collector terminal and returned through V_{CC} to the emitter. I_{EB} represents the few electrons that do not reach the collector junction but exit through the base terminal and return via V_{BE} to the emitter terminal.

From Fig. 4-6(b) and (c) we see that the relationships between terminal

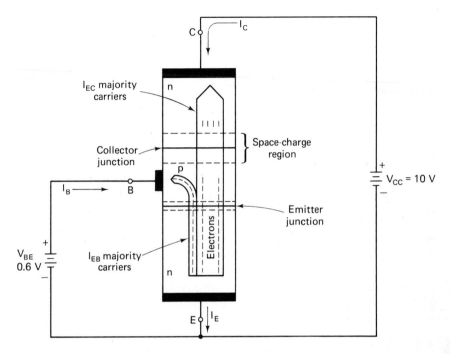

(a) npn internal transistor currents

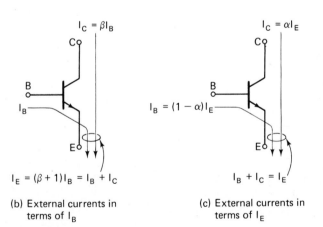

(b) External currents in
terms of I_B

(c) External currents in
terms of I_E

Figure 4-6 *npn* transistor action. External currents are shown as conventional to agree with the transistor symbol's arrowhead.

currents I_B, I_C, and I_E are the same for the *npn* BJT as they were for the *pnp* BJT. Transistor electrical characteristics α and β still represent how I_E splits into I_B and I_C, so all the equations and examples in Section 4-5 apply to Fig. 4-6. These principles are illustrated in the next three examples.

Example 4-6: In Fig. 4-7(a), find (a) type of transistor, *npn* or *pnp*; (b) direction of I_E; (c) direction of I_C; and (d) direction of I_B.
Solution: (a) Since the emitter arrow points toward the emitter terminal, the emitter is *n*-type and the BJT is an *npn*. (b) I_E is downward to agree with the emitter arrowhead. (c) I_C flows left, resulting from conventional current flow *out* of the (+) terminal of V_{CC}. (d) I_B flows right, resulting from conventional current flow out of the (+) terminal of V_{BE}.

Example 4-7: Repeat the analysis of Example 4-6 for Fig. 4-7(b).
Solution: (a) Arrow points toward *n*-type base, identifying a *pnp* BJT. (b) I_E is upward. (c) I_C is right, flowing into the (−) terminal of V_{CC}. (d) I_B is left, flowing into the (−) terminal of V_{BE}.

Once the direction of I_E was established as downward in part (b) of Example 4-6, we could also have reasoned that both I_B and I_C must flow into the transistor terminals. That is, current leaving the BJT I_E must equal current entering the BJT, or $I_B + I_C$. The same reasoning could be applied after part (b) of Example 4-7 to conclude that I_B and I_C flow out of the BJT terminals.

Example 4-8: In Fig. 4-7(a) and (b), find the *magnitude* of I_B and I_E.
Solution: From Eq. (4-2),

$$I_B = \frac{I_C}{\beta} = \frac{990 \ \mu A}{99} = 10 \ \mu A$$

From Eq. (4-1), $I_E = 10 \ \mu A + 990 \ \mu A = 1000 \ \mu A = 1mA$. As a check,

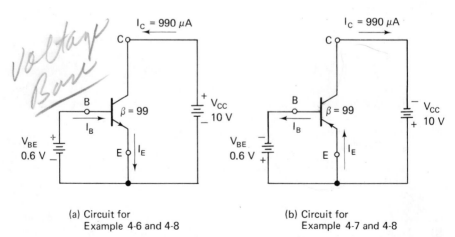

(a) Circuit for
Example 4-6 and 4-8

(b) Circuit for
Example 4-7 and 4-8

Figure 4-7 Circuits for Examples 4-6 to 4-8.

use Eq. (4-5):

$$I_E = (\beta + 1)I_B = (99 + 1)(10 \ \mu A) = 1000 \ \mu A = 1 \text{ mA}$$

4-7 Preview of Biasing

4-7.1 Need for Biasing. The BJT is most commonly used to amplify signals. A signal is defined as a variation in voltage produced by a signal generator. For example, the playback head of a tape recorder generates signals in the order of 20 mV according to the pattern of information recorded on the tape. A speaker may require 10 V across its terminals to reproduce the taped sound at a comfortable hearing level. An amplifier circuit which consists of BJTs, resistors, and capacitors is therefore required between the tape head and the speaker to amplify, to 10 V, the 20-mV signal.

Before the BJT can amplify a signal, its junctions must be properly biased (emitter junction forward-biased and collector junction reverse-biased) to establish the BJT in a linear or active operating mode. As noted in Section 4-2, the emitter junction must be forward-biased and the collector junction reverse-biased. It is true that the circuits of Figs. 4-3, 4-4, 4-6(a), and 4-7 do establish a linear operating mode for the BJT. These circuits are very useful to develop the basic ideas of transistor action. However, they must be changed into a more practical biasing arrangement, for reasons detailed in the next section.

4-7.2 Base-Current Control. In Fig. 4-7 dc base-bias current I_B is set (for illustration purposes only) by battery V_{BE}. Now if we wanted to set I_B at 10 μA as in Examples 4-6 to 4-8, we would first have to get, somewhere, extremely accurate data on exactly what value of V_{BE} is needed for *this particular* transistor to establish $I_B = 10 \ \mu A$. The required V_{BE} might be 0.580 V. And let's assume, again for this transistor, that $V_{BE} = 0.600$ V will give an I_B of 20 μA and $V_{BE} = 0.575$ V will give an I_B of 5 μA. Clearly it would be an impossible job to buy a battery that would stay at precisely 0.580 V; and we would need an expensive, accurate voltmeter to check the battery. So, controlling V_{BE} is not the answer when trying to bias the BJT.

The answer is to control base current by installing a base-bias resistor R_B, as in Fig. 4-8. R_B should be as large as possible so that its $I_B R_B$ voltage drop will be large with respect to V_{BE}. Thus, if I_B must be 10 μA and we pick $R_B = 400 \text{ k}\Omega$, the $I_B R_B$ drop would be

$$I_B R_B = V_{RB} = (10 \times 10^{-6} \text{ A})(400 \times 10^3 \ \Omega) = 4.0 \text{ V}$$

Base supply battery V_{BB} must feed the drop across both R_B and V_{BE}, or

$$V_{BB} = I_B R_B + V_{BE} \qquad (4\text{-}8)$$

So V_{BB} must be set at 4.6 V from

$$V_{BB} = 4.0 + 0.6 = 4.6\ V$$

Now if V_{BB} or V_{BE} vary by a few tenths of a volt, I_B will stay relatively constant. This principle is explored in the next example to show that Fig. 4-8 is an example of *constant-base-current biasing*. Note in Fig. 4-8 that V_{BE} is only the voltage from base to emitter.

Example 4-9: In Fig. 4-8, $V_{BB} = 4.60$ V and $R_B = 400$ kΩ. Find I_B if V_{BE} equals (a) 0.575 V, (b) 0.600 V, and (c) 0.580 V.

Solution: Rewrite Eq. (4-8) in terms of I_B:

$$I_B = \frac{V_{BB} - V_{BE}}{R_B} = \frac{4.6 - V_{BE}}{0.4 \times 10^6}$$

(a) $I_B = \dfrac{4.600 - 0.575}{400{,}000} = 10.06\ \mu A$

(b) $I_B = \dfrac{4.600 - 0.600}{400{,}000} = 10.00\ \mu A$

(c) $I_B = \dfrac{4.600 - 0.580}{400{,}000} = 10.04\ \mu A$

Example 4-7 shows that variations in V_{BE} (also V_{BB}) of a few tenths of a volt have a negligible effect on I_B. We can draw a general conclusion from Fig. 4-8 and Example 4-9 that *BJTs are current-controlled rather than voltage-controlled devices.*

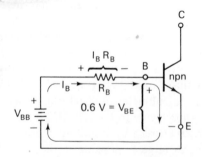

Figure 4-8 Control of base-bias current.

4-7.3 Load Resistor. In the introductory circuits of Figs. 4-3 through 4-7, no provision was made for a load. That is an element whose power would be controlled by the transistor. Loads such as voice coil speakers, other transistor circuits, lamps, and motors are easily modeled by a resistor. The resistor that represents or models the load is appropriately called a *load resistor* and is identified on the circuit schematic by the symbol R_L.

As shown in Fig. 4-9, the load resistor is located in series with either

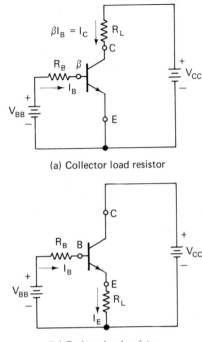

(a) Collector load resistor

(b) Emitter load resistor

Figure 4-9 Location of load resistors.

collector or emitter terminal. Load resistors are rarely, if ever, located in series with the base terminal, because base currents are so small that far less current can be delivered to the load than can be delivered by either collector or emitter current.

Introduction of R_L means that collector–emitter voltage V_{CE} can vary, since V_{CC} no longer is connected across the E and C terminals of the BJT. V_{CC} must now divide between R_L and V_{CE} as in Fig. 4-9. We shall analyze fully in Chapters 5 and 6 the manner in which V_{CC} divides.

4-8 AC Circuit Configurations

4-8.1 Introduction. Assuming that a BJT is properly biased into its active mode of operation, two questions arise. (1) How can we connect a weak voltage signal from a signal generator as an *input* to the BJT circuit? (2) How can we extract an amplified reproduction of the signal as an *output* from the BJT circuit? The answer is that there are six possible ways of accomplishing these tasks.

Figure 4-10 illustrates all six possible combinations. Each combination is called a *circuit configuration*, and each circuit configuration has a name. The

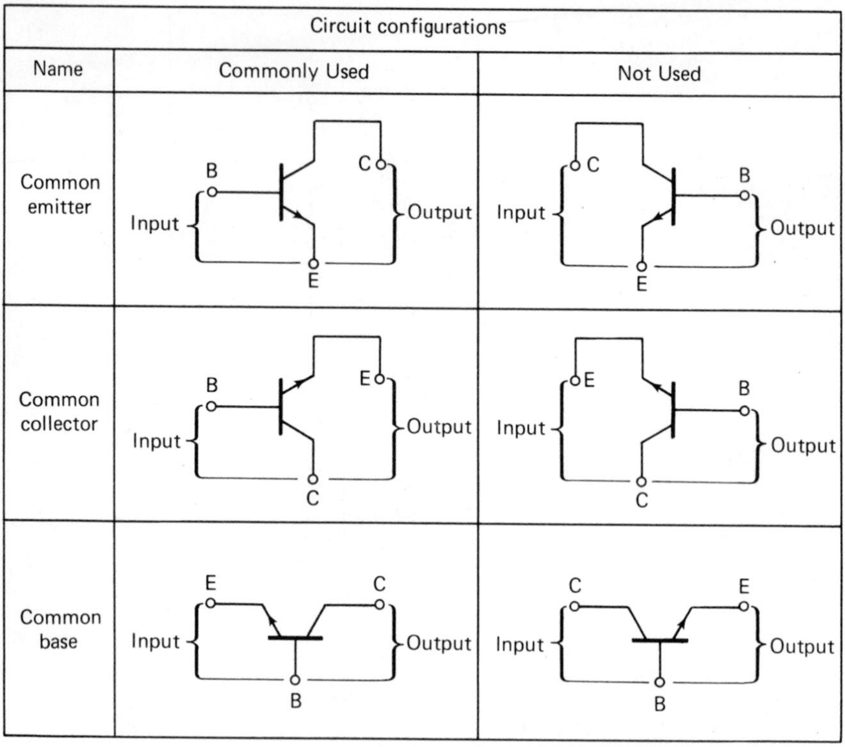

Circuit configurations		
Name	Commonly Used	Not Used
Common emitter		
Common collector		
Common base		

Figure 4-10 Input and output connection possibilities for a BJT.

name is defined by the transistor terminal that is connected to or *common* to both the input and the output. The names are, in order of importance, common emitter (abbreviated CE), common collector (abbreviated CC), common base (abbreviated CB). Although all configurations in Fig. 4-10 are drawn for *npn* transistors, the same terminology is used for *pnp* transistors.

The left-hand column gives the three practical configurations of these that are commonly used. Those configurations in the right-hand column are not used because they do not provide amplification. Note that the signal generator is never connected to collector terminal *C*. Also the output is never taken at the base terminal *B*. In conclusion, refer to the left-hand column, where the input currents flow into either the base, *B*, or emitter, *E*, terminals. The output currents flow from either the collector, *C*, or emitter, *E*, terminals.

4-8.2 Simplified Common-Emitter, Common-Collector, and Common-Base Circuit Configurations. Bias voltages and resistor R_B are omitted in Fig. 4-11 so as to focus attention strictly on the methods of connecting input signal E_i to a BJT and extracting output signal V_D. The omissions also make it easier to identify the common BJT terminal.

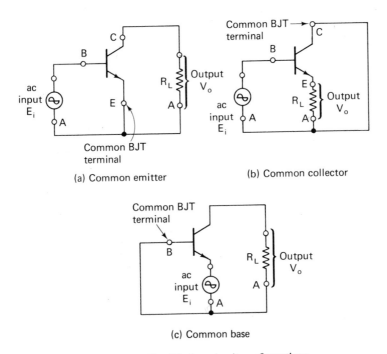

(a) Common emitter

(b) Common collector

(c) Common base

Figure 4-11 Simplified ac circuit configurations.

In Fig. 4-11(a), emitter terminal E_i is common to terminal A of E_i and terminal A of R_L. This method of connection identifies a common-emitter circuit. A common-collector circuit is defined in Fig. 4-11(b) because collector terminal C is common to both E_i and R_L at their respective A terminals. Finally, in Fig. 4-11(c), the A terminal of E_i and the A terminal of R_L are common to base terminal B. This ac circuit arrangement specifies a common-base circuit.

Figure 4-11 emphasizes that circuit configuration is concerned only with how ac input E_i and ac output V_o are obtained with a BJT. The dc mode of operation is active for each ac circuit configuration.

At this time, it is natural to ask how the circuit configurations are determined for actual circuits. This will be examined next.

4-8.3 Practical Common-Emitter, Common-Collector, and Common-Base Circuit Configurations. Dc bias supply voltages V_{BB} and V_{CC}, together with R_B, are shown in all three ac circuit configurations of Fig. 4-12 to establish all circuits in the linear operating mode. The main ideas that must next be applied are that V_{BB}, V_{CC}, and all capacitors in Fig. 4-12 have negligible reactance or resistance to the flow of ac signal currents. Again, V_{BB}, V_{CC}, and all capacitors are ac short circuits and provide a short circuit between their terminals to the flow of ac currents. As a matter of fact, if we model V_{CC}, V_{BB},

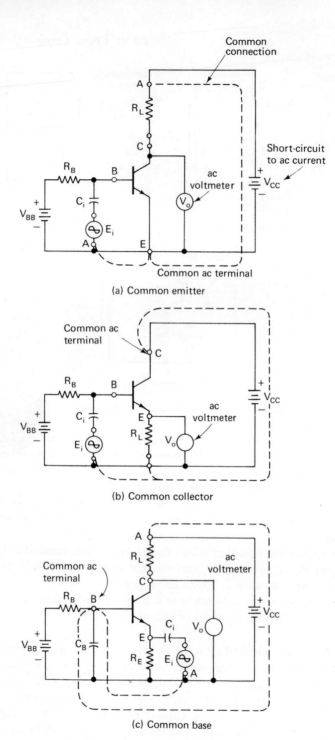

(a) Common emitter

(b) Common collector

(c) Common base

Figure 4-12 Practical ac circuit configurations.

and each capacitor by a short circuit, Fig. 4-12 will be simplified to Fig. 4-11.

In Fig. 4-12(a), E_i is connected through ac short circuit C_i between B and E (C_i also prevents V_{BB} from forcing a dc current through E_i). R_L is connected through ac short circuit V_{CC} between C and E. Therefore, E is common to both E_i and R_L (A terminals) to identify a common-emitter circuit.

C_i connects E_i between B and C in Fig. 4-12(b). C_i and V_{CC} provide the ac short circuit shown by the dashed line. R_L is connected between E and C, so Fig. 4-12(b) specifies a common-emitter circuit.

In Fig. 4-12(c), capacitor C_B places an ac short across both R_B and V_{BB}. Resistor R_E is added both to prevent a short circuit around E_i and to provide a path for dc emitter current I_E. As shown by the dashed lines, E_i is connected via C_B and C_i between E and B. R_L is connected between C and B by V_{CC} and C_B. Base terminal B is common to both E_i and R_L (A terminals) and identifies a common-base circuit. Before we study each circuit configuration in detail in the next two chapters, it is well to have a general idea of their relative performance capabilities.

4-8.4 Performance Comparison Among Common-Emitter, Common-Collector, and Common-Base Circuit Configurations.

When analyzing or measuring the ac performance of a BJT amplifier we are usually interested in answering one or more of the following questions:

1. If we apply an ac input voltage, E_i, how much ac output voltage, V_o, will we obtain across load R_L? It is more common to obtain the ratio of V_o to E_i and specify the ratio V_o/E_i as *voltage gain*.
2. What resistance is presented by the BJT to the flow of signal currents from E_i? The answer is, *input resistance*.
3. What is the lowest signal frequency that can be amplified by the BJT and still give a useful output? The answer is, *low-frequency response*.
4. *High-frequency response* is the answer to the question: What is the highest signal frequency that can be amplified by the BJT and still give a useful output?

A general comparison of these performance criteria is given in Table 4-2.

Table 4-2 PERFORMANCE COMPARISON OF CIRCUIT CONFIGURATIONS

| Circuit | Gain | | Resistance | | Frequency Response | |
| | | | | | Low | High |
Configuration	Current	Voltage	Input	Output	Cutoff	Cutoff
CE	High	High	Medium	High	Medium	Medium
CC	High	$\simeq 1$	High	Low	Excellent	Excellent
CB	$\simeq 1$	High	Low	High	Poor	Excellent

Problems

4-1 Draw the circuit symbol for a *pnp* and *npn* BJT.

4-2 On the circuit symbol for an *npn* BJT, does the arrowhead point toward the base or the emitter?

4-3 Does the BJT arrowhead point in the direction of conventional or electron current flow?

4-4 Name the operating modes and state the required bias condition on each junction.

4-5 What is the operating mode of a BJT amplifier?

4-6 If V_{BE} is reversed in Fig. 4-3, what type of bias exists at each junction (forward or reversed) and what operating mode results?

4-7 Should current enter the collector or the emitter terminal of a *npn* BJT in the linear mode?

4-8 I_E measures 1000 μA in Fig. 4-4 and I_B measures 10 μA. Find I_C.

4-9 Which transistor parameter, α or β, gives the ratio of collector current to base current?

4-10 An obsolete data sheet shows $\alpha = 0.990$. Find β.

4-11 What changes occur in Example 4-1 when a BJT is substituted with $\beta = 75$?

4-12 Find I_B and I_C if $I_E = 1$ mA and $\beta = 99$.

4-13 I_B is set for 5 μA in a linear circuit. The BJT has $\beta = 99$. Find I_C and I_E.

4-14 If 1000 electrons flow from the emitter to the base region, how many reach the collector terminal if $\alpha = 0.995$?

4-15 Would the BJT of Problem 4-14 be an *npn* or a *pnp*?

4-16 Is the BJT in Fig. 4-7 an *npn* or a *pnp*?

4-17 To forward-bias the base–emitter junction of an *npn* transistor, what polarity of voltage must be applied to the base with respect to the emitter?

4-18 Should the collector terminal of an *npn* transistor be made positive or negative for linear operation?

4-19 What changes result in Example 4-8 if $\beta = 49$?

4-20 What base currents flow in Example 4-9 if $V_{BB} = 9.0$ V?

4-21 What are the most common BJT circuit configurations used in the active mode of operation?

4-22 Between which terminals is the input signal applied for a CE amplifier?

4-23 What circuit element is used to model the behavior of coupling capacitors and supply voltages with respect to ac voltages and currents?

4-24 What performance characteristics of a BJT amplifier are of most importance?

4-25 What effect do the capacitors in Fig. 4-12 have on the dc BJT currents?

5

Common-Emitter DC Circuit Analysis

5-0 Introduction

A preview of biasing was given in Chapter 4 to show, very briefly, how a BJT circuit is operated in the linear mode. Recall that *biasing* is defined as that procedure whereby emitter and collector junctions are properly biased to place the BJT in the required operating mode. In this chapter we will study biasing in more detail, but our studies will concentrate on biasing the BJT into the *linear mode* of operation. Keep firmly in mind that biasing is concerned with establishing proper values for dc currents through the BJT. We shall *not* attempt to connect an ac signal input to the BJT until Chapters 5 and 6. This separation of dc circuit analysis from ac circuit analysis is quite deliberate and very practical. When trouble shooting a BJT circuit, it is accepted practice to turn off all signals and first measure dc voltages or currents in the circuit. For if dc bias currents are incorrect, it is futile to check ac signals. Furthermore, a simultaneous analysis of both ac and dc circuit operations is difficult, especially during early encounters with BJT circuits. Separate analysis of dc and ac performance simplifies our problems.

Before delving into the details of biasing we need to know more about the relationships between current I through two BJT terminals and voltage V across the same two terminals. For convenience the current–voltage relationships are plotted as an I–V characteristic and are furnished by the manufacturer. Since the most popular BJT circuit is the common emitter configuration, we concentrate on its characteristic curves.

5-1 Common-Emitter Input *I–V* Characteristic

5-1.1 Measuring the I_B–V_{BE} Characteristic Curve. As was shown in Section 4-8.2, base and emitter terminals are the input pair for a CE circuit. Its input characteristic curve is therefore a plot of input current I_B versus input voltage V_{BE}. We can measure this I_B – V_{BE} characteristic with the circuit of Fig. 5-1(b). First set $V_{CC} = 0$. This ensures that $V_{CB} = 0$ V. Then increase V_{BB} from 0 to about 6 V in convenient steps. At each step, measure corresponding values of base current I_B, with dc microammeter A, and base–emitter voltage V_{BE} with dc VTVM, V. The value of I_B that corresponds with V_{BE} is plotted as one point. For example, in Fig. 5-1(b) point A shows that $I_B =$

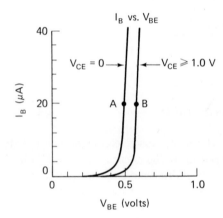

(a) Input I_B-V_{BE} characteristic curves

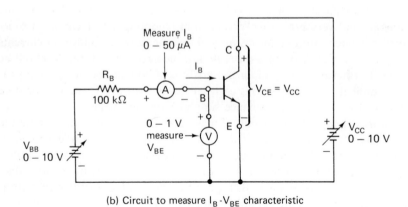

(b) Circuit to measure I_B-V_{BE} characteristic

Figure 5-1 CE input I_B–V_{BE} characteristic curves for an *npn* BJT. (Reverse all battery and meter connections for a *pnp*.)

20 μA when $V_{BE} = 0.5$ V. When all points are connected, we obtain the I_B–V_{BE} characteristic corresponding to $V_{CE} = 0$ in Fig. 5-1(a). As we should have expected, this curve has the same general shape as a diode curve. It is, in fact, the characteristic of the base–emitter diode.

Now increase V_{CC} to 10 V in Fig. 5-1(b). V_{CE} will increase from zero volts so that collector C goes positive with respect to emitter E. As long as V_{CE} exceeds about 1.0 V, the collector junction is reverse-biased, and collector current flows through the emitter junction. Then we increase V_{BB} from zero volts to vary I_B in convenient steps. As before, each value of I_B, for example 20 μA, has a corresponding value of V_{BE}, for example 0.6 V. These corresponding values are plotted as point B in Fig. 5-1(a). When sufficient points are plotted and connected, we obtain the I_B–V_{BE} characteristic, labeled $V_{CE} \geq 1.0$ V in Fig. 5-1(a) ($\geq$ means equal to or greater than).

5-1.2 Using the I_B–V_{BE} Characteristic Curve. Recall that when a BJT is used in an amplifier circuit, it must operate in the linear mode (Section 4-2). This means that its collector junction must be reverse-biased, and consequently V_{CB} should exceed about 1.0 V. So the I_B–V_{BE} input characteristic curve that we use is the one labeled $V_{CE} \geq 1.0$ V in Fig. 5-1(a).

The most important fact to be learned from this curve is that $V_{BE} \simeq 0.6$ V for almost all values of dc base current I_B. So, as long as the base–emitter terminals are forward-biased, they behave like a battery whose value is approximately 0.6 V. This conclusion applies only for dc base currents. For a germanium transistor we would obtain curves of the same general shape except that (1) the I_B–V_{BE} curve for $V_{CE} = 0$ would approximate the vertical line $V_{BE} = 0.2$ V, and (2) the I_B–V_{BE} curve for $V_{CE} \geq 1.0$ V would approximate the vertical line $V_{BE} \simeq 0.3$ V. We conclude that we will use $V_{BE} = 0.6$ V for a silicon transistor and $V_{BE} = 0.3$ V for a germanium transistor in all biasing problems. We use these conclusions in the next two examples.

Example 5-1: Calculate dc base bias current I_B for a silicon BJT in Fig. 5-2(a). *Solution:* Since we have a silicon BJT, we approximate V_{BE} as 0.6 V. V_{BB} must supply the voltage drop across R_B or $I_B R_B$ and also voltage drop V_{BE}. Expressed as an equation,

$$V_{BB} = I_B R_B + V_{BE} \tag{5-1a}$$

Solving for I_B,

$$I_B = \frac{V_{BB} - V_{BE}}{R_B} \tag{5-1b}$$

Substituting values into Eq. (5-1b),

$$I_B = \frac{(6 - 0.6)\ \text{V}}{100{,}000\ \Omega} = 54\ \mu\text{A}$$

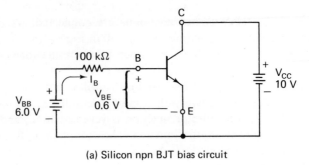

(a) Silicon npn BJT bias circuit

(b) Germanium pnp BJT bias circuit

Figure 5-2 Circuits for Examples 5-1 and 5-2.

Example 5-2: Calculate dc bias current I_B for the germanium BJT in Fig. 5-2(b).

Solution: Choose $V_{BE} = 0.3$ V with the polarity shown in Fig. 5-2(b). Substituting values into Eq. (5-1b),

$$I_B = \frac{V_{BB} - V_{BE}}{R_B} = \frac{(6.0 - 0.3) \text{ V}}{100,000 \text{ }\Omega} = 57\mu\text{A}$$

By comparing Example 5-1 with Example 5-2 we can learn that as long as V_{BB} is large with respect to V_{BE} ($V_{BB} \geq 3$ V), the value of V_{BE} has very little effect on I_B. That is, substituting a germanium for a silicon transistor made a difference of only 3 μA in I_B.

5-2 Common-Emitter Collector Characteristic Curves

5-2.1 Defining the I_C–V_{CE} Characteristic. In a CE circuit configuration the collector terminal furnishes output current to the load. Since the common terminal is the emitter, output voltage from the BJT is collector-to-emitter voltage V_{CE}. Therefore, the required output characteristic is an I_C–V_{CE} characteristic curve. We might be tempted to vary a battery voltage across collector and emitter and measure the resulting collector current. But we learned from our study of transistor action in Section 4-3 that base current also controls collector current. It is meaningless to vary both I_B and V_{CE} at the same time and record values for I_C. Someone else would vary I_B and V_{CE} in a different manner and inevitably obtain different results. The accepted method of presenting an I_C–V_{CE} characteristic will be developed from the method of obtaining its measurement in Section 5-2.2. Characteristics of only silicon BJTs will be considered.

5-2.2 Measuring the I_C–V_{CE} Characteristic Curve in the Cutoff Mode.

In the test circuit of Fig. 5-3(a), adjust V_{BB} to zero volts to set I_B equal to zero. This ensures that the emitter junction is *not* forward-biased and prevents majority carriers from being injected into the base and crossing the collector junction. Any currents due to minority carriers crossing the emitter junction are so small that collector current is essentially zero (less than 0.5 μA).

Having now established $I_B = 0$, we increase V_{CC} to adjust V_{CE} from 0 to 10 V. Ammeter A_2 will read 0 mA throughout the change in V_{CE}. If we plot $I_C = 0$ for each value of V_{CE} in Fig. 5-3(b), we obtain an I_C–V_{CE} characteristic curve that is a straight line and lies on top of the horizontal V_{CE} axis. From a review of Section 4-2 we conclude that this curve identifies the *cutoff mode of operation* (both emitter and collector junctions are reverse-biased).

Figure 5-3(b) gives us a better visualization of the cutoff mode. The transistor is cut off in the sense that no current is conducted between C and E, so that the C and E terminals act as an open circuit.

5-2.3 Measuring the I_C–V_{CE} Characteristic Curve in the Saturation Mode.

In Fig. 5-3(a) V_{CC} is set to zero volts to ensure that $V_{CE} = 0$ V. Next raise V_{BB} to 2.6 V to fix I_B at 20 μA. I_B will now stay constant. Let V_{CE} be increased slowly to about 0.2 V. Milliammeter A_2 would show that I_C increases as V_{CE} increases. If we plotted corresponding values of I_C and V_{CE} we would obtain the almost straight line, in Fig. 5-3(b), that rises from the origin to point 1. There are three basic ideas that we must think of when we see this type of curve.

First, point 1 in Fig. 5-3(b) is not mere dot on a paper. Any point on a characteristic curve has a specific name, called an *operating point*. Point 1 tells us that $I_C = 0.9$ mA, $V_{CE} = 0.2$ V, and $I_B = 20$ μA when the BJT is biased at this operating point.

Second, we can evaluate all BJT terminal voltages to prove that point 1 identifies an operating point in the saturation mode. We can assume that $V_{BE} = 0.6$ V, because the emitter junction is forward-biased. Showing the polarities and values of V_{BE} and V_{CE} in Fig. 5-4(a), we conclude that V_{CB}

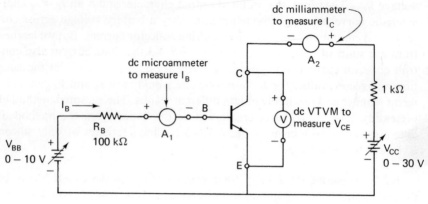

(a) Circuit to measure an I_C-V_{CE} characteristic curve

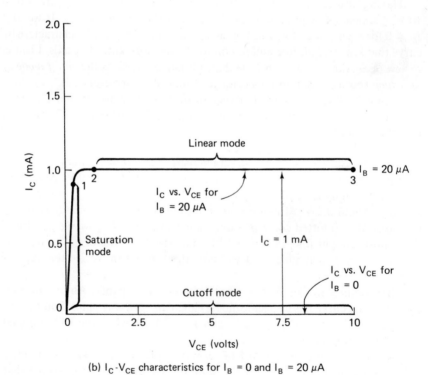

(b) I_C-V_{CE} characteristics for $I_B = 0$ and $I_B = 20 \ \mu A$

Figure 5-3 Measurement of I_C–V_{CE} characteristic curves for a silicon BJT.

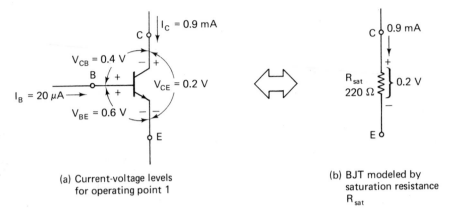

(a) Current-voltage levels
for operating point 1

(b) BJT modeled by
saturation resistance
R_{sat}

Figure 5-4 BJT saturation-mode operation from point 1 in Fig. 5-3(b).

must have the polarity shown and a value of 0.4 V. Note that V_{CB} forward-biases the collector junction. Since both junctions are forward-biased, we find from Section 4-2 that the BJT is biased into its saturated mode.

Third, all points on the curve from origin to point 1 trace out the saturation mode of operation where I_C increases directly with V_{CE}. Recall that voltage across a resistor increases directly with current through the resistor. So the BJT in the saturated mode (only) behaves like a resistor. The ratio of V_{CE} to I_C (in the saturation mode only) will tell what value of resistance is presented by the BJT between its C and E terminals. This resistance has a special name, *saturation resistance* (abbreviated R_{sat}) R_{sat} is modeled in Fig. 5-4(b) and is calculated from

$$R_{sat} = \frac{V_{CE}}{I_C} \quad \text{in saturation mode}$$

Example 5-3: Calculate R_{sat} for operating point 1 in Fig. 5-3(b).
Solution:

$$R_{sat} = \frac{V_{CE}}{I_C} = \frac{0.2 \text{ V}}{0.9 \text{ mA}} \simeq 200 \text{ } \Omega$$

5-2.4 Interpreting the I_C–V_{CE} Characteristic Curve in the Linear Mode. In Fig. 5-3(b), refer to the I_C–V_{CE} characteristic curve, labeled $I_B = 20 \text{ } \mu A$. The portion of this curve that lies between points 2 and 3 contains all possible operation points for operating the BJT in its linear mode. We should realize that by knowing that a BJT is biased for operation at point 3, we immediately know what its terminal currents and voltages are. Conversely, if we want to operate a BJT at point 3, we can tell what currents and voltages must be delivered to it by the biasing circuit. As an example, we assume that $V_{BE} \simeq$

0.6 V (forward bias) because current $I_B = 20$ μA [see point B in Fig. 5-1(a)]. Next, point 3 lies on the horizontal line that intersects $I_C = 1$ mA. So collector current equals 1 mA, and I_E can be calculated from Eq. (4-1) as

$$I_E = I_C + I_B = 1 \text{ mA} + 20 \text{ } \mu\text{A} = 1020 \text{ } \mu\text{A}$$

Next, point 3 lies on a vertical line extending down to intercept the horizontal axis at $V_{CE} = 10$ V.

All data for operating point 3 are shown on the BJT symbol in Fig. 5-5. Observe that V_{CB} is clearly equal to 9.4 V and reverse-biases the *npn*'s collector junction.

Example 5-4: It is desired to operate a BJT at operating point 2 in Fig. 5-3(b). What values of current and voltages must be delivered to the BJT by the circuit?

Solution: Since $I_B = 20$ μA, $V_{BE} = 0.6$ V. The value of I_C is 1 mA, identical to I_C for point 3. Unlike point 3, $V_{CE} = 1.0$ V at point 1. The operating-point data would be identical with Fig. 5-5 except that $V_{CE} = 1.0$ V and $V_{CB} = 0.4$ V.

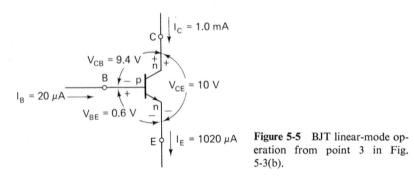

Figure 5-5 BJT linear-mode operation from point 3 in Fig. 5-3(b).

5-3 Beta Measurement from I_C–V_{CE} Characteristic Curves

5-3.1 Family of I_C–V_{CE} Characteristic Curves. In Section 5-2 and Fig. 5-3 we analyzed, in detail, two I_C–V_{CE} characteristic curves. It is standard practice for transistor manufacturers to give performance data on their typical BJTs as a *family* of I_C–V_{CE} curves. V_{CE} is varied over its maximum operating range, typically between 0 and 25 V, for each convenient value of base current. The result is as shown in Fig. 5-6.

There are commercial curve plotters or curve tracers that display the entire family of characteristic curves. A step-current generator holds base current constant for 8.3 ms while V_{CE} is swept (varied) over the desired range. The step-current generator then abruptly steps base current up to the next desired

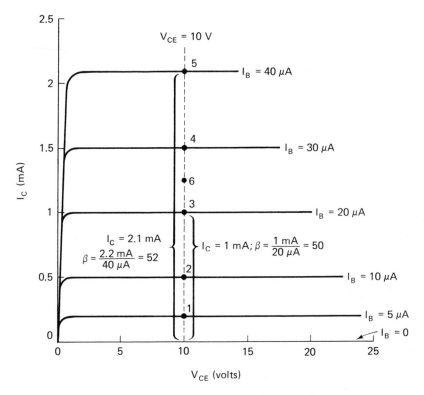

Figure 5-6 Family of I_C–V_{CE} collector characteristic curves.

constant value and V_{CE} is again swept. Normally between 4 and 12 base-current steps can be selected to display a family of 4 to 12 characteristic curves. If available, operating a curve plotter is the best way of testing and learning about transistors.

5-3.2 Measurement of Beta. Transistor beta was defined in Section 4-3.3 and Eq. (4-2) as the ratio of collector current I_C to base current I_B (in the linear mode), or

$$\beta = \frac{I_C}{I_B}$$

There are two methods of presenting the relation between I_C and I_B for a particular type of BJT. Either the manufacturer specifies values for β as h_{FE} on his data sheet, or he gives a plot of collector characteristic curves, as in Fig. 5-6. Since both methods present the same data, we should be able to find β from an I_C–V_{CE} characteristic and also draw the I_C–V_{CE} characteristic if β is known. We will learn from the next example how to calculate β from the collector characteristics.

Example 5-5: Calculate β for operating points 1 through 6 in Fig. 5-6. Note that $V_{CE} = 10$ V for each operating point.

Solution: Referring to operating point 5 in Fig. 5-6, note first that it is in the active mode since $V_{CE} \geq 1$ V. Next, point 5 lies on the curve $I_B = 40\ \mu A$. Finally a horizontal projection from point 5 to the I_C axis shows that $I_C = 2.1$ mA. Substituting into Eq. (4-2) gives

$$\beta = \frac{2.1\ \text{mA}}{40\ \mu A} = \frac{2100\ \mu A}{40\ \mu A} = 52.5$$

It is possible that an operating point does not lie on one of the characteristics. In this event estimate what I_B line it would lie on. For example, point 6 is halfway between $I_B = 20\ \mu A$ and $I_B = 30\ \mu A$, so point 6 probably lies on the line $I_B = 25\ \mu A$ and corresponds to $I_C = 1.25$ mA. Thus for point 6, $\beta = 1.25\ \text{mA}/25\ \mu A = 50$. Calculations for all operating points are tabulated for comparison in Table 5-1.

Table 5-1 β FOR POINTS 1–5

Point	I_B (μA)	I_C (mA)	$\beta = I_C/I_B$
1	5	0.20	40
2	10	0.50	50
3	20	1.00	50
6	25	1.25	50
4	30	1.50	50
5	40	2.10	52

5-4 Using Beta and the I_C–V_{CE} Characteristic

We conclude from Example 5-5 that β can vary with operating-point collector current. Furthermore, manufacturers give β for their typical transistors, plus minimum and maximum values. For example, a BJT with typical β of 100 could have a spread between a minimum of 30 and a maximum of 400. For purposes of design or analysis, use the typical value. If you have a curve plotter and the BJT, its β can easily be measured. If no curve plotter is available, β of the BJT can be calculated from measurements of I_C and I_B obtained either from the working BJT circuit or from the test circuit of Fig. 5-3(a). These comments are amplified in the next examples.

Example 5-6: If $I_B = 40\ \mu A$ and $\beta = 100$, what value of I_C results, assuming the BJT is in the linear mode?

Solution: From Eq. (4-2), $I_C = \beta\ I_B = 100 \times 40\ \mu A = 4$ mA.

Example 5-7: If we need $I_C = 1$ mA and $\beta = 100$, what value of I_B will be required for the BJT?

Solution: From Eq. (4-2), $I_B = I_C/\beta = 1\ \text{mA}/100 = 10\ \mu A$.

Example 5-8: Our BJT has characteristics like those of Fig. 5-6. It operates in a linear circuit that required $I_C = 1.5$ mA. Do we need to know β to find the required value of base bias current I_B?

Solution: On Fig. 5-6, locate $I_C = 1.5$ mA on the I_C axis and find the closest I_C-V_{CE} linear characteristic curve that corresponds with the horizontal line $I_C = 1.5$ mA. We locate the curve $I_B = 30$ μA (point 4). Since we now know that I_B must be 30 μA, we do not need the value of β.

5-5 Biasing with a Single Battery for Constant Base Current

5-5.1 Single-Battery Biasing Circuit. The biasing circuits introduced in Section 4-7 have a major disadvantage: they all require two dc power supplies, one for the base circuit and another for the collector circuit. It is far more practical and economical to arrange the bias circuit so that the collector power supply also supplies bias current to the base. In Fig. 5-7(a) a single dc supply V_{CC} will supply collector current I_C to load R_L and the BJT. By adding a single base-bias resistor R_B, V_{CC} also supplies base-bias current I_B.

5-5.2 Constant-Base-Bias Current. The base-bias current loop is shown in Fig. 5-7(a) and again in Fig. 5-7(b), where it is clear that V_{CC} supplies both V_{BE} and the voltage drop across R_B, according to

$$V_{CC} = I_B R_B + V_{BE} \qquad (5\text{-}2a)$$

Often V_{BE} is small with respect to V_{CC} ($V_{BE} \ll V_{CC}$) and, as discussed in Section 4-7.2, base-bias current I_B will be held constant by the constant-current source V_{CC} and R_B. That is, by neglecting V_{BE} in Eq. (5-2a), I_B depends only on R_B and V_{CC} from

$$I_B \cong \frac{V_{CC}}{R_B}, \qquad V_{BE} \ll V_{CC} \qquad (5\text{-}2b)$$

As will be shown in Example 5-13 and Section 5-7.4, constant-base-bias current biasing does have a disadvantage.

5-5.3 Collector Loop. The complete loop current circuit for I_C defines the output loop for a CE circuit in Fig. 5-7(a). The output loop is redrawn in Fig. 5-7(b) to emphasize the fact that V_{CC} divides between R_L and the BJT collector and emitter terminals according to

$$V_{CC} = I_C R_L + V_{CE} \qquad (5\text{-}3a)$$

where voltage drop V_{RL} across R_L is expressed by

$$V_{RL} = I_C R_L \qquad (5\text{-}3b)$$

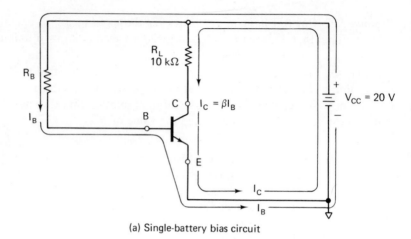

(a) Single-battery bias circuit

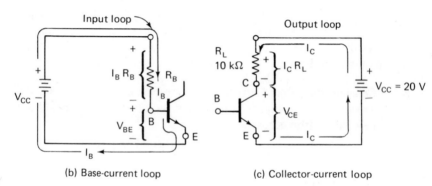

(b) Base-current loop   (c) Collector-current loop

Figure 5-7 Single-battery bias circuit in (a) has a constant-base-current loop in (b). V_{CC} divides between $I_C R_L$ and V_{CE} in the output loop of (c).

5-6 DC Analysis of the Single-Battery Constant-Base-Current Bias Circuit

5-6.1 Analysis of the Collector Loop. Dc analysis of any bias circuit usually begins with the output loop. The first point to consider is what limits are placed on V_{CE} by the circuit, if the BJT is to stay biased into the active mode. This point is discussed in the next example.

Example 5-9: In Fig. 5-7(a) or (c), what are the (a) minimum and (b) maximum allowable values of V_{CE} to hold the BJT in its active mode?
Solution:

(a) V_{CE} should at least equal and preferably exceed 1 V in order to reverse bias the collector junction and keep the BJT out of the saturation mode.

(b) V_{CC} determines the maximum voltage that can be developed across V_{CE}. If I_C equaled 0 mA, there would be zero volts dropped across R_L, and all of V_{CC} would develop across V_{CE}. Of course, when $I_C = 0$, the BJT is in the cutoff mode, so the maximum value of V_{CE} is slightly less than 20 V and the allowable range for V_{CE} is

$$V_{CE} = 1 \text{ V to } V_{CE} \leq V_{CC} = 20 \text{ V}$$

In the next examples we find the allowable range for I_C to hold the BJT in its active mode.

Example 5-10: Find the corresponding values of I_C for the (a) minimum V_{CE} and (b) maximum V_{CE} found in Example 5-9.
Solution:

(a) From Example 5-9(a), minimum V_{CE} is 1.0 V. From Eq. (5-3),

$$I_C = \frac{V_{RL}}{R_L} = \frac{V_{CC} - V_{CE}}{R_L} = \frac{(20 - 1) \text{ V}}{10 \text{ k}\Omega} = 1.9 \text{ mA}$$

(b) From Example 5-9(b), the maximum limit on V_{CE} is 20 V, so, from Eq. (5-3),

$$I_C = \frac{V_{RL}}{R_L} = \frac{V_{CC} - V_{CE}}{R_L} = \frac{(20 - 20) \text{ V}}{10 \text{ k}} = 0 \text{ mA}$$

The maximum range for I_C is 0 to 1.9 mA. Note that maximum I_C occurs at minimum V_{CE} and vice versa.

Example 5-11: It is desired to set V_{CE} equal to $\frac{1}{2}V_{CC}$ or 10 V in Fig. 5-7. What are the required collector bias current and V_{RL} for this operating point?
Solution: From Eq. (5-3a),

$$I_C = \frac{V_{CC} - V_{CE}}{R_L} = \frac{(20 - 10) \text{ V}}{10 \text{ k}\Omega} = 1 \text{ mA}$$

From Eq. (5-3b), $V_{RL} = I_C R_L = 1 \text{ mA} \times 10 \text{ k}\Omega = 10 \text{ V}$.

Analysis of the collector loop usually results in a determination of collector-current value. To find the corresponding base current we need (see Section 5-4) to know β of the BJT or to have a graph of its characteristic curves. Thus we make the transition from collector (output) circuit to base (input) circuit through β, as will now be shown.

5-6.2 Analysis of the Base-Current Loop. In Example 5-11 we found that to operate the BJT of Fig. 5-7 at $V_{CE} = 10$ V we need I_C equal to 1 mA. What do we do next to make $I_C = 1$ mA? V_{CC} and R_L are fixed, so there is nothing we can adjust in the output. What we must do is find what value of I_B will set I_C equal to 1 mA and then install the correct value of R_B to set the required value of I_B. The procedure is illustrated by the following example.

Example 5-12: Make $V_{CE} = 10$ V and $I_C = 1$ mA in the circuit of Fig. 5-7. Assume that $\beta = 50$.
Solution: Find the required I_B from Eq. (4-2):

$$I_B = \frac{I_C}{\beta} = \frac{1 \text{ mA}}{50} = 20 \ \mu\text{A}$$

Observe also that point 3 in Fig. 5-6 corresponds exactly with the operating point of this problem and that β is also identical. Therefore, from Fig. 5-6, $I_B = 20 \ \mu\text{A}$. Next, refer to Fig. 5-7(a) or (b) and Eq. (5-2b) to find R_B:

$$R_B \cong \frac{V_{CC}}{I_B} = \frac{20 \text{ V}}{20 \times 10^{-6} \text{ A}} = 1 \text{ M}\Omega$$

5-6.3 Disadvantages of Constant-Base-Current Biasing. The constant-base-current bias circuit works fine if we do not change anything once the circuit has been set up properly. For example, assume that we have built the circuit of Examples 5-11 and 5-12 and obtained an operating point of $I_B = 20 \ \mu\text{A}$, $I_C = 1$ mA, and $V_{CE} = 10$ V for BJT with a $\beta = 50$. These operating data are shown in Fig. 5-8(a). Now, if our original transistor is replaced with one that has a different β, we want to learn what changes will occur in the circuit. The changes will be examined by means of an example.

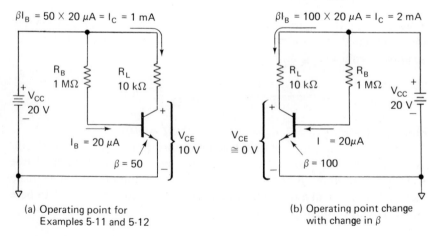

(a) Operating point for
   Examples 5-11 and 5-12

(b) Operating point change
   with change in β

Figure 5-8 Effect of β variation on the operating point with constant-base-current bias.

Example 5-13: The BJT in Fig. 5-8(a) has $\beta = 50$. What happens if it is replaced with a BJT that has $\beta = 100$?

Solution: I_B will be unchanged at 20 μA, because $V_{BE} \approx 0.6$ V no matter what the value of β. However, from Eq. (4-2), I_C will increase to $I_C = \beta I_B = 100 \times 20 \mu$A $= 2$ mA. Now V_{RL} will rise to 2 mA $\times$ 10 k$\Omega = 20$ V and will take all of V_{CC}. This condition is shown in Fig. 5-8(b), where the BJT is deep into its saturation mode, with V_{CE} equal to a few tenths of a volt, or approximately zero volts.

We conclude from Example 5-13 that collector-operating-point current I_C will vary with changes in β as long as base-bias current is held constant. Before presenting, in Chapter 6, bias circuits that hold collector currents fairly constant with changes in β, it will be productive to review these basic ideas from another viewpoint.

The ideas developed thus far in Sections 5-5 and 5-6 may be difficult to grasp at first. Fortunately, a graphical technique is available as a visual aid that shows how the BJT and external circuit work together. This technique is circuit analysis by means of *load lines*. It is presented next.

5-7 Graphical Analysis of Circuits with Constant-Base-Bias Current

5-7.1 Introduction to the BJT DC Load Line. We can superimpose the collector-loop *external* circuit R_L and V_{CC} [see Fig. 5-7(c)] on top of the BJT's collector characteristics by a graphical technique known as drawing the dc load line. This technique will be illustrated with respect to Figs. 5-7 and 5-8. Recall from Example 5-10 that I_C could only range between about 0 mA and $V_{CC}/R_L = 20$ V/10 k$\Omega = 2$ mA. It is important to emphasize that this range for I_C depends *not* on the BJT but on R_L and V_{CC}. We therefore need a graph with an I_C axis that extends from 0 to V_{CC}/R_L and a V_{CE} axis that extends from 0 to V_{CC}. Next we shall plot a dc load line on this graph, initially *without* paying any attention to any of the I_C-V_{CE} characteristics. Equation (5-3a) is actually the equation of the dc load line and is repeated here for convenience:

$$V_{CC} = I_C R_L + V_{CE} \tag{5-3a}$$

Every point on the dc load line shows one possible combination of I_C and V_{CE} that is allowed by the *circuit*, that is, R_L and V_{CC}. Each point on a transistor's I_C-V_{CE} characteristic is a possible combination of I_C and V_{CE} that is allowed by the transistor. It follows that the intersection of load line and I_C-V_{CE} characteristic is the only point that simultaneously is allowed by *both* circuit and transistor. This point of intersection is the familiar *operating point* and will be identified by the symbol O. Finally, the particular $I_C\text{–}V_{CE}$ characteristic curve that the BJT operates on is determined by I_B.

This graphical analysis technique, which finds the circuit and BJT operating point, is summarized and amplified as follows:

1. Pretend that a short circuit exists between C and E so that $V_{CE} = 0$ in both the circuit and Eq. (5-3a). Solve for the maximum possible I_C (ignoring saturation voltage) from

$$I_{Cmax} = \frac{V_{CC}}{R_L}, \qquad \text{at } V_{CE} = 0 \qquad (5\text{-}4)$$

2. Pretend that an open circuit exists between C and E so that $I_C = 0$ in both circuit and Eq. (5-3a). Solve for the maximum possible V_{CE} from

$$V_{CEmax} = V_{CC}, \qquad \text{at } I_C = 0 \qquad (5\text{-}5)$$

3. Obtain a graph of characteristic curves for the BJT. Draw a straight line from the maximum current point $V_{CE} = 0$, $I_C = V_{CC}/R_L$ to the minimum current point at, $V_{CE} = V_{CC}$, $I_C = 0$. This line is the dc load line.
4. From the circuit, evaluate I_B, for example $I_B = 20\ \mu A$, and follow it to its intersection with the load line. The intersection is labeled point O and its coordinates tell the value of I_B, V_{CE}, and I_C that exist in both BJT and circuit. The graphical-analysis technique will be illustrated in the next section.

5-7.2 Drawing the DC Load Line. In Examples 5-10 through 5-12 we analyzed thoroughly the circuit of Fig. 5-8(a). We shall analyze this circuit by load-line techniques (so that our results can be verified) in the next example.

Example 5-14: In the circuit of Fig. 5-9(a), the BJT has characteristics as given in Fig. 5-6. Draw the dc load line.
Solution: As instructed in the summary of Section 5-7.1:

1. From Eq. (5-3a), at $V_{CE} = 0$, $I_C = V_{CC}/R_L = 20\text{ V}/10\text{ k}\Omega = 2\text{ mA}$.
2. From Eq. (5-3a), at $I_C = 0$, $V_{CE} = V_{CC} = 20\text{ V}$.
3. A graph of the BJT $I_C\text{–}V_{CE}$ characteristics in the range of steps 1 and 2 is given in Fig. 5-9(b).

The maximum current point from step 1 is shown on the I_C axis at 2 mA and the minimum current point from step 2 is shown on the V_{CE} axis at 20 V. Connecting these points gives the dc load line. Note that we have not used step 4 yet because we only drew the load line.

It should be pointed out that any other point on the load line can be plotted if it is more convenient than plotting end points for the load line. To illustrate, pick any convenient value of V_{CE} from inspection of the V_{CE} axis, for example 5 V. Now substitute this value for V_{CE} into Eq. (5-3a).

$$V_{CC} = I_C R_L + V_{CE}$$
$$20 = I_C \times 10,000 + 5$$

Solve for I_C:

$$I_C = \frac{15}{10,000} = 1.5 \text{ mA}$$

Now plot this point $V_{CE} = 5$ V, $I_C = 1.5$ mA as point A on Fig. 5-9(b). Since any two points determine a straight line, we could connect point A with a straightedge to either end point and extend the line to the remaining end point to complete the load line.

Example 5-15: Finish the graphical analysis begun in Example 5-14 by locating operating point O for the circuit in Fig. 5-9(a).

Solution: From Eq. (5-2b), evaluate I_B.

$$I_B \simeq \frac{V_{CC}}{R_B} = \frac{20 \text{ V}}{1 \text{ M}\Omega} = 20 \text{ } \mu\text{A}$$

From step 4 of the summary in Section 5-7.1, locate the I_C–V_{CE} characteristic curve in Fig. 5-9(b), labeled 20 μA. Its intersection with the load line is shown by operating point O. The vertical dashed line extending down from point O intersects the V_{CE} axis at 10 V. The drawing below the V_{CE} axis pictures how V_{CC} divides between V_{CE} and R_L.

In conclusion, graphical analysis has located the circuit's operating point. Point O in Fig. 5-9 tells us the value of V_{CE}, I_C, and I_B, and with these values we can easily calculate other BJT and circuit behavior. For example, $V_{CE} = 10$ V, $I_C = 1$ mA, and $I_B = 20$ μA for point O. Therefore, $\beta = I_C/I_B = 1$ mA, $V_{RL} = I_C R_L = 1$ mA $\times$ 10 k$\Omega = 10$ V.

5-7.3 Saturation and Cutoff on the Load Line. The load line gives a clear picture of BJT saturation and cutoff. In Fig. 5-10 the load line for Fig. 5-9 is redrawn for convenience. If base current is reduced to zero, we find the $I_B = 0$ curve in Fig. 5-10 and locate its intersection with the load line at point C. Here $I_C = 0$ and $V_{CE} = V_{CC}$, so point C locates cutoff-mode operation.

If I_B is now increased slowly to 20 μA, the operating point travels up the load line to point O. If I_B is further increased to 40 μA or more, the operating point travels up the load line and stops at point S. Note that operating point S is the same operating point for both $I_B = 40$ μA and $I_B = 50$ μA. This

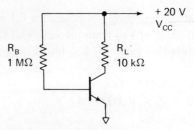

(a) Circuit for Examples 5-14 and 5-15

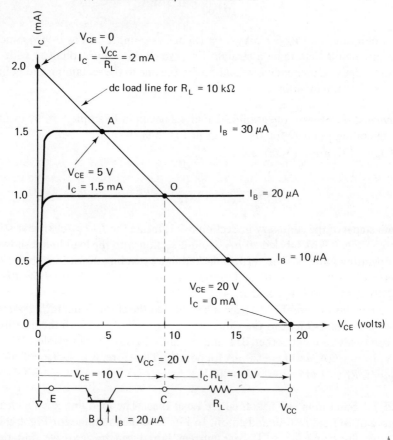

(b) Load line and operating point

Figure 5-9 The circuit in (a) is analyzed by drawing the dc load line in (b).

means that the BJT is conducting all possible current ($I_{C\max} \approx V_{CC}/R_L$) allowed by the circuit. In other words, the BJT is operating in its saturated mode and point S identifies saturation. Operating points on the load line between points S and C are in the linear mode, as shown.

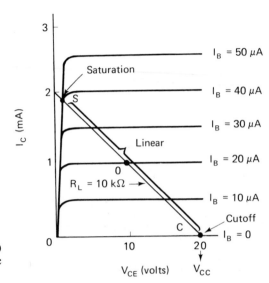

Figure 5-10 BJT saturation(s) and cutoff (c) points on the dc load line.

We have studied the effects of varying I_B on the operating point. All that remains is to study the change in operating point because of changes in R_L, V_{CC}, and β.

5-7.4 Effect on the Operating Point of Varying R_L, V_{CC}, and β. Base current is held constant in both circuits of Fig. 5-11. In Fig. 5-11(a) V_{CC} is held constant at 20 V. With $R_L = 0$ the load line is a vertical line and the operating point is located at point 1. As R_L is increased to 10 kΩ, the operating point travels along the characteristic curve labeled $I_B = 20 \ \mu A$ to point 2. With $R_L = 20$ kΩ, the BJT enters saturation at point 3. As R_L is further increased, it remains in saturation until cutoff, when R_L approaches an infinitely large value.

In Fig. 5-11(b) R_L is held constant at 10 kΩ. As V_{CC} is reduced from 20 V to 10 V and then to zero, the operating point moves along the characteristic $I_B = 20 \ \mu A$ from linear operation at point 1 into saturation at point 2 and then to cutoff at the origin.

The effect on operating-point location for changes in β was investigated in Example 5-13. The same effect is graphically portrayed by comparison of Fig. 5-12(a) and (b). In both, I_B is held constant at 20 μA and the circuits are identical, with $V_{CC} = 20$ V and $R_L = 10$ kΩ. Both load lines must also be the same. In Fig. 5-12(a), operating point O is located at $V_{CE} = 10$ V, $I_C = 1$ mA. If this BJT, which has $\beta = 50$, is replaced with a BJT that has $\beta = 100$, the operating point moves into saturation, as shown in Fig. 5-12(b). This dramatic shift in operating-point location with change in β shows the disadvantage of a bias circuit that holds the base current constant. In the next

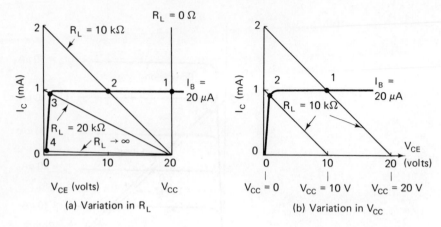

Figure 5-11 Effect of varying R_L in (a) or V_{CC} in (b) on the operating point.

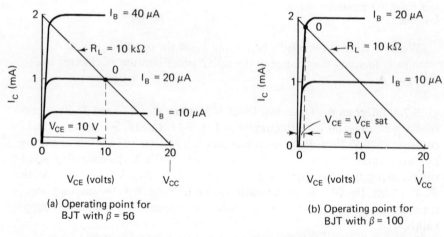

Figure 5-12 Effect of a change in β on the operating point.

chapter we shall analyze a circuit that holds collector current constant. We will also learn how the operating point shifts along the load line as the result of a signal voltage to generate an amplified output voltage across a load.

Problems

5-1 What are the input terminals for a common-emitter circuit?

5-2 What are typical values of base-to-emitter voltage under operating conditions for a (a) silicon transistor; (b) germanium transistor?

5-3 If V_{BB} in Fig. 5-2 is changed to 4 V, what is base current I_B? Assume use of a silicon transistor.

5-4 Repeat Problem 5-3 for a germanium transistor.

5-5 What are the two output terminals of a common-emitter circuit?

5-6 When a transistor is in the cutoff mode, are the emitter and collector junctions forward- or reverse-biased?

5-7 When both the emitter and collector junctions are forward-biased, what is the mode of operation?

5-8 What element does a BJT behave like in the saturation mode?

5-9 Calculate the saturation resistance of a BJT if $V_{CEsat} = 0.4$ V and $I_C = 2$ mA.

5-10 What mode of operation must a BJT be biased to operate as an amplifier?

5-11 What are the operating-point values (I_B, I_C, V_{CE}) for point 3 in Fig. 5-3?

5-12 Calculate β at operating point 3 in Fig. 5-3.

5-13 A transistor operating in the linear mode has $I_B = 30$ μA and $\beta = 60$; calculate dc collector current I_C.

5-14 From measured values at the operating point, $I_B = 50$ μA and $I_C = 8$ mA, calculate β.

5-15 Refer to Fig. 5-7(a); if $R_B = 2$ MΩ, calculate (a) base current, I_B; (b) collector current, I_C. $\beta = 80$.

5-16 Let $I_B = 10$ μA and $\beta = 60$ in Fig. 5-7(a) and calculate the (a) voltage drop across R_L; (b) voltage drop from emitter to collector.

5-17 If $R_L = 5$ kΩ in Fig. 5-9(a), (a) draw the load line in Fig. 5-9(b); (b) locate the operating point.

5-18 If $V_{CC} = 15$ V in Fig. 5-9(a), (a) draw the load line; (b) locate the operating point. $R_L = 10$ kΩ and $R_B = 1$ MΩ.

6

Common-Emitter
Large-Signal Operation

6-0 Introduction

We extend the load-line technique introduced in Chapter 5 to show graphically how input signal currents and voltages are processed by a BJT. We will also learn how operating point location is stabilized by adding a resistor in series with the emitter.

How does a transistor circuit produce amplification of a signal, commonly called voltage gain? What factors control the maximum possible voltage that can be delivered to a load? What happens to V_{CE} when collector current is varied by a signal? These questions will be answered in this chapter. But first we must differentiate between a large signal (Chapter 6) and a small signal (Chapter 7).

6-1 Wave-form Terminology

Up to this point we have been concerned only with preparing the BJT circuit to process a signal. This preparation process was called biasing or establishing a dc operating point in the active mode. The net result was to bias the BJT with proper dc base and collector currents and dc voltages V_{CE} and V_{RL}.

Voltages from a signal source, for example tape head, phone cartridge, or antenna, are ac in nature rather than dc. It is much more convenient for us to use a signal generator to test or analyze BJT circuits, because the signal generator's amplitude and frequency can be controlled precisely. A signal generator's output usually has a sinusoidal wave form, as shown in Fig. 6-1a, where

E_{ip} = peak value of the sinusoidal voltage as measured by a cathode-ray oscilloscope

$E_i = 0.707E_{ip}$ = rms signal voltage as measured with an ac voltmeter

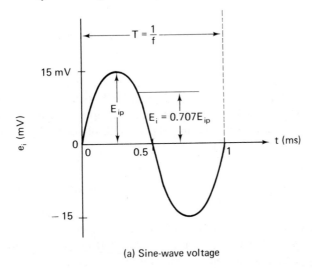

(a) Sine-wave voltage

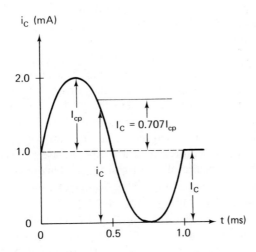

(b) Sine-wave signal current I_{cp} superimposed on dc bias current I_C

Figure 6-1 Signal voltage in (a) causes signal current I_{cp} to vary around dc bias current I_C in (b).

t = time, usually in milliseconds (ms) or microseconds (μs)
T = period or time for one complete signal cycle
f = frequency (in hertz), the number of cycles completed in 1 s

Period, T, and frequency, f, are related by

$$T = \frac{1}{f} \quad \text{or} \quad f = \frac{1}{T} \tag{6-1}$$

Many waveforms encountered in a BJT circuit are made up of an ac signal combined with a dc bias. For example, in Fig. 6-1(b) the collector current at any instant, i_C, has an average or dc value, I_C, on which is superimposed a sinusoidal variation in collector current with a peak value of I_{cp}. I_C is caused by the dc bias circuitry; I_{cp} is caused by the introduction of signal voltage E_i. Note our use of the simplifying principle of superposition. Think first of the dc component (biasing). Then think of the ac signal component riding or superimposed on the dc component.

6-2 Graphical Analysis of Large-Signal Operation

6-2.1 Base Currents—DC and AC. Dc bias currents are shown as dashed lines in the common-emitter circuit of Fig. 6-2. [Review Fig. 5-8(a) and (b).] Input voltage E_i is a sinusoidal signal source with a peak value of $E_{ip} = 15$ mV or peak-to-peak value of $E_{ip/p} = 30$ mV. Coupling capacitor C_C is required to block the flow of dc current through E_i. Without C_C, the current through R_B would flow into E_i and not into the base, to drive the operating point down along the load line into cutoff.

The frequency of E_i is 1000 Hz [see Fig. 6-1(a)]. In one cycle E_i goes through zero volts three times and no signal base current will flow, as shown in Fig. 6-2(a). The peak value of base signal current is $I_{bp} = 10$ μA and occurs at $t = 0.25$ ms and $t = 0.75$ ms, where the peak values of $E_{ip} = 15$ mV. On its positive half-cycle, E_i pumps current into the base as in Fig. 6-2(b).

We will not use a graphical technique to show why a peak value of precisely $E_i = 15$ mV is required to cause a peak base current of 10 μA. Their ratio would be a measure of input resistance. This would be an exercise in futility, because a base–emitter voltage change of 15 mV is impossible to plot around a base–emitter bias voltage of 0.6 V. A simple, direct method of evaluating I_b will be shown in Chapter 7. For now we have established that E_i causes a sinusoidal signal current to be coupled by C_c into the base.

Total base current at any instant, i_B, is the sum of dc bias current I_B and ac signal current I_b. At time zero, $t = 0$, $I_B = 20$ μA and base signal current is zero. Net base current is $i_B = I_B = 20$ μA in Fig. 6-2(a). The operating-point location is shown as point O in Fig. 6-3.

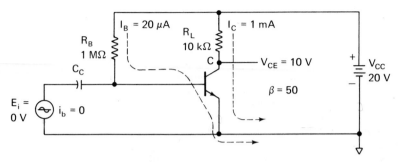

(a) dc bias currents, t = 0, 0.5 ms and
   1.0 ms, operating point O

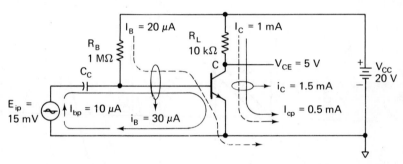

(b) Positive peak signal current plus
    bias current, t = 0.25 ms, point A

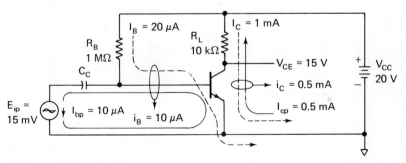

(c) Negative peak signal current plus
    bias current, t = 0.75 ms, point B

Figure 6-2   Circuit conditions for operating points in Fig. 6-3.

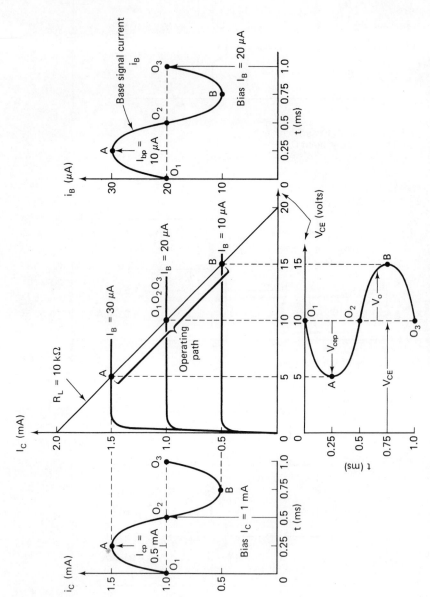

Figure 6-3. Graphical analysis of signal currents and voltage for the common-emitter circuit of Fig. 6-2.

6-2.2 Operating Path. At $t = 0.75$ ms, we see from Fig. 6-2(b) that total base current i_B is the sum of $I_B = 20 \ \mu A$ and $I_{bp} = 10 \ \mu A$ or $i_B = 10 \ \mu A$ because $I_{bp} = 10 \ \mu A$ subtracts from $I_B = 20 \ \mu A$. We can now plot the total base current variation, i_B, over one cycle, as shown to the right of the load line in Fig. 6-3.

At time $t = 0.25$ ms, $i_B = 30 \ \mu A$, so the new operating point will be located at the intersection of load line and characteristic curve $I_B = 30 \ \mu A$. This intersection is labeled point A in Fig. 6-3. At time $t = 0.75$ ms, $i_B = 10 \ \mu A$ and the operating point is located at point B on the load line. We conclude that as i_B varies over one complete cycle, from O_1 to A to O_2 to B to O_3, the operating point traces a path along the dc load line from O to A to O to B and back to O. The path AOB is appropriately called the *operating path*.

6-2.3 Collector Currents—DC and AC. As the operating point moves along the operating path in Fig. 6-3, collector current i_C and collector–emitter voltage must change in step. For example, point A on the load line occurs at $t = 0.25$ ms, where i_C has increased to 1.5 mA and v_{CE} has decreased to 5 V. At point B on the load line ($t = 0.75$ ms) $i_C = 0.5$ mA and $v_{CE} = 15$ V. As shown to the left of the characteristic curves in Fig. 6-3, the variation in i_C can be plotted against time. The ac component of collector current is a sine wave with a peak value $I_{cp} = 0.5$ mA that rides on the dc bias component $I_C = 1.0$ mA. Points A and B of i_C correspond to points A and B on both load line and i_B.

6-2.4 Collector and Output Signal Voltages. By dropping vertical projections from points O, A, and B in Fig. 6-3, the shape of collector voltage v_{CE} can be plotted against time. Note carefully the sequence of cause and effect from base current to collector voltage. At point A, a signal current I_{bp} of $10 \ \mu A$ causes a collector current of $I_{cp} = 0.5$ mA. Their relationship will be examined in Section 6-4. The *change* in collector current I_{cp} causes a *change* in voltage drop across R_L that is equal to V_{cep}, or

$$V_{cep} = I_{cp}R_L \qquad (6\text{-}2a)$$

Now recall that signal input voltage E_{ip} was a sine wave that (1) caused I_{bp} to (2) cause I_{cp} to cause V_{cep}. Thus the sine wave with peak value V_{cep} is the end result of the BJT circuit amplifying E_{ip}. So V_{cep} is really the ac *signal output voltage* V_o, whose peak value is designated V_{op}. Therefore,

$$V_{op} = I_{cp}R_L \qquad (6\text{-}2b)$$

The relationship between V_o and E_i will be explored further in the next section.

6-3 Voltage Gain

In Fig. 6-4 we focus attention on just the input and output ac signal voltages developed in Figs. 6-2 and 6-3. Capacitor C_o is added in Fig. 6-4 to block the dc component of v_{CE}. Only the ac component of v_{CE} is coupled through C_o, to appear as output voltage V_o. The ratio of V_o to E_i is called *voltage gain* and has the symbol A_v. However, we see that during the first half-cycle, when E_i goes positive, V_o goes negative. Voltage polarities at instant A are shown in Fig. 6-4 for both E_i and V_o. Clearly the output goes negative (with respect to the common grounded emitter) when the input goes positive. This means that input and output are out of phase by 180°. In this text we shall account for the phase difference by reference to the circuit schematic and will *not* include a minus sign in any equation to account for phase reversal. Therefore, the magnitude of V_o, voltage gain, is given by

$$A_V = \frac{V_o}{E_i} \tag{6-3}$$

Example 6-1: Evaluate the voltage gain of the circuit in Fig. 6-4.
Solution: The peak value of E_i is 15 mV, and the corresponding peak value of V_o is 5 V. From Eq. (6-3),

$$A_v = \frac{V_o}{E_i} = \frac{5}{0.015} = 333$$

Example 6-2: If E_i was reduced to 1 mV peak to peak, what would be the peak-to-peak output voltage in Example 6-1?

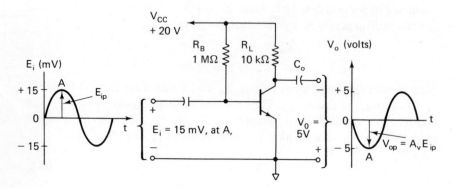

Figure 6-4 Common-emitter voltage gain.

Solution: Since A_v was evaluated as 333 in Example 6-1, we use Eq. (6-3) to find V_o from

$$V_{op/p} = A_v E_{ip/p} = 333 \times 1\,\text{mV} = 333\,\text{mV}$$

6-4 AC Beta

We now examine the relation between the *change* in collector current and *change* in base current that was observed in Section 6-2.3. Recall that in Section 4-3.3 the relationship between dc collector current, I_C, and dc base current, I_B, was defined by Eq. (4-2) as

$$\beta = \frac{I_C}{I_B}$$

It helped to think of I_B as causing a larger collector current equal to βI_B. The same analogy can be used to think of a change in I_B, or ΔI_B (Δ means change), as causing a change in I_C, or ΔI_C.

Often the spacings between the horizontal portions of the base current lines on the I_C–V_{CE} curves are unequal, as in Fig. 6-5. This means that the ratio of dc currents I_C/I_B may not equal the ratio of ac current $\Delta I_C/\Delta I_B$. To differentiate between the two ratios, we could give each ratio a different name and symbol. For example, the ratio I_C/I_B could be called β_{dc}. The ratio of

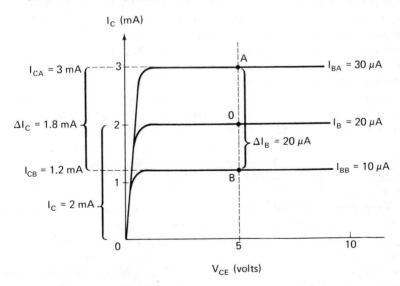

Figure 6-5 Ac $\beta = \Delta I_C/\Delta I_B$ may differ from dc β because of unequal spacing between I_C–V_{CE} curves.

$\Delta I_C / \Delta I_B$ could be called β_{ac}. These relationships are given next in equation form together with the symbol used on some data sheets:

$$h_{FE} = \beta_{dc} = \frac{I_C}{I_B} = \text{dc beta} \tag{6-4a}$$

$$h_{fe} = \beta_{ac} = \frac{\Delta I_C}{\Delta I_B} = \text{ac beta, } V_{CE} \text{ held constant} \tag{6-4b}$$

Their differences are explored in an example.

Example 6-3: A BJT has collector characteristics given in Fig. 6-5. (a) Evaluate β_{dc} at operating point 0. (b) Evaluate β_{ac} and ac collector current for a base-current swing of 10 μA peak around the operating point.
Solution:

(a) From Fig. 6-5, $I_B = 20$ μA and $I_C = 2.0$ mA at dc operating point O. From Eq. (6-4a),

$$\beta_{dc} = \frac{I_C}{I_B} = \frac{2.0 \text{ mA}}{20 \text{ } \mu\text{A}} = \frac{2 \times 10^{-3}}{20 \times 10^{-6}} = 100$$

(b) Since I_{bp} is given as 10 μA, I_B will vary between limits of $I_{BA} = I_B + I_{bp}$ $= 30$ μA and $I_{BB} = I_B - I_{bp} = 10$ μA. For purposes of uniformity, the corresponding limits of collector current are measured on the peak I_B lines, I_{BA} and I_{BB}, directly above and below the operating point. We construct a vertical dashed construction line through O to intersect the peak I_B lines at points A and B in Fig. 6-5. It is simpler and more accurate to use peak-to-peak measurements on a graph. Therefore, peak-to-peak ΔI_B is measured from I_{BA} to I_{BB} or $\Delta I_B = 30$ μA $- 10$ μA $= 20$ μA. Point A locates $I_{CA} = 3.0$ mA and point B locates $I_{CB} = 1.2$ mA. The corresponding peak-to-peak change in I_C is $I_{CA} - I_{CB} = 3.0$ mA $- 1.2$ mA or $\Delta I_C = 1.8$ mA. From Eq. (6-4b),

$$\beta_{ac} = \frac{\Delta I_C}{\Delta I_B} = \frac{1.8 \text{ mA}}{20 \text{ } \mu\text{A}} = \frac{1.8 \times 10^{-3}}{20 \times 10^{-6}} = 90$$

Very often, magnitudes of β_{ac} and β_{dc} are within 10% of being equal. Furthermore, we often have available only approximate values for either β_{ac} or β_{dc}. Therefore, in most practical circuits and throughout the rest of this text we will assume that $\beta_{ac} = \beta_{dc}$ and *use the simpler symbol β for both.* Specifically, β will be evaluated from Eq. (4-2) as I_C / I_B and used interchangeably for ac or dc beta.

Where convenient, rms or average peak values may be used for peak-to-peak values. Peak values I_{cp} and I_{bp} are equal to half the p/p values. Substituting into Eq. (6-4b),

$$\beta_{ac} = \frac{I_{cp}}{I_{bp}} = \frac{\Delta I_C/2}{\Delta I_B/2} = \frac{0.9 \text{ mA}}{10 \ \mu A} = 90 \qquad (6\text{-}4c)$$

For root-mean-square (rms) values I_c and I_b, $I_c = 0.707 I_{cp}$ and $I_b = 0.707 I_{bp}$

$$\beta_{ac} = \frac{I_C}{I_B} = \frac{0.707 I_{cp}}{0.707 I_{bp}} = \frac{0.63 \text{ mA}}{7.0 \ \mu A} = 90 \qquad (6\text{-}4d)$$

6-5 Maximum Swing in Output Voltage

In Section 5-7.1 the minimum and maximum limits for V_{CE} were given by Eqs. (5-4) and (5-5). The lower limit for V_{CE} was idealized at zero volts when the transistor was driven into saturation. Actually, minimum V_{CE} is between 0.1 and 0.5 V, corresponding to the BJT's saturation voltage. Maximum V_{CE} occurs when the BJT is cut off and equals V_{CC}. In Section 6-2.4 it was shown that output voltage V_o was equal to the variation in v_{CE}. By biasing the operating point at the center of the load line, as in Fig. 6-6(a), the maximum possible *symmetrical* peak swing can be obtained for V_o. It is approximately equal to one-half of V_{CC}.

Suppose that the operating point is located on the upper half of the load line, as in Fig. 6-6(b). The maximum V_{op} that can be obtained without distortion equals the bias value of V_{CE}. That is, operating point O can be driven along the load line to saturation. If E_i and I_b are increased further, no change occurs in v_{CE} or V_o. So changes in E_i will not be reproduced in V_o, and distortion results. Note that point O could be driven farther down the load line into cutoff. But it is the *smaller* of the distances from O to cutoff and O to saturation that determines maximum V_{op}.

In Fig. 6-6(c) operating point O is located on the lower half of the load line. The smaller distance that O can move is into cutoff. The corresponding change in collector current equals bias current I_C, and from Eq. (6-2) the resulting maximum output voltage is $I_C R_L$.

Two general conclusions can be drawn concerning maximum output voltage:

1. Locate the operating point at the center of the load line for maximum possible output voltage. Its maximum V_{op} will equal $V_{CC}/2$.
2. If the operating point is not located at the center of the load line, maximum peak-output voltage will equal the smaller of (a) operating-point voltage V_{CE} or (b) the product of operating-point current I_C and R_L.

Example 6-4: In a CE circuit $V_{CC} = 24$ V, $R_L = 6$ kΩ, $\beta = 50$, and the

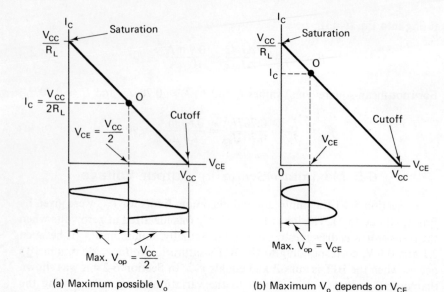

(a) Maximum possible V_o

(b) Maximum V_o depends on V_{CE}

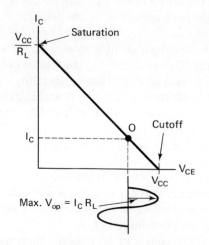

(c) Maximum V_o depends on $I_C R_L$

Figure 6-6 Dependence of maximum output voltage on operating-point location.

operating point is located at $I_C = 1$ mA, $V_{CE} = 18$ V. (a) Find the maximum available peak output voltage V_{op}. What is the required (b) peak ac collector current I_{cp} and (c) peak ac base current I_{bp} to obtain V_{op} in part (a)? *Solution:* (a) Apply conclusion 2 above. Evaluate $I_C R_L = 1$ mA $\times$ 6 k$\Omega = 1 \times 10^{-3} \times 6 \times 10^3 = 6$ V. Compare it with $V_{CE} = 18$ V and conclude that the maximum allowable peak output voltage is the smaller of $I_C R_L$ or V_{CE}; that is, $V_{op} = 6$ V. (b) By reference to Fig. 6-6(c) we conclude that I_{cp} must swing through a value equal to I_C or $I_{cp} = I_C = 1$ mA. (c) We employ Eq. (6-4b) but use $\beta = 50$ for β_{ac} and $I_{cp} = 1$ mA for ΔI_C. Then solve for $\Delta I_B = I_{bp}$:

$$\beta = \frac{I_{cp}}{I_{bp}} \quad \text{or} \quad I_{bp} = \frac{I_{cp}}{\beta} = \frac{1 \text{ mA}}{50} = 20 \; \mu A$$

Example 6-5: In the circuit of Example 6-4, where should the operating point be located for maximum possible peak output voltage?
Solution: From conclusion 1,

$$V_{CE} = \frac{V_{CC}}{2} = \frac{24 \text{ V}}{2} = 12 \text{ V}$$

Find I_C from Eq. (5-3a):

$$I_C = \frac{V_{CC} - V_{CE}}{R_L} = \frac{(24 - 12) \text{ V}}{6000} = 2 \text{ mA}$$

From Eq. (4-2), $I_B = I_C/\beta = 2$ mA/50 $= 40 \; \mu A$.

The operating point must be located at $I_C = 2$ mA, $V_{CE} = 12$ V, $I_B = 40 \; \mu A$, and the peak possible output voltage will be 12 V.

6-6 AC Load

6-6.1 Modeling the Actual Load. Up to now the load has been modeled by a single load resistor, R_L. In every circuit R_L has conducted a dc bias current at all times. Then, when an ac signal was applied to the input, an ac current component, I_{cp}, was superimposed upon the dc current.

Loads, such as a speaker, require only the ac current component. Dc current through a speaker will offset the cone, to cause distortion and probably damage. It may also be necessary to couple only the ac signal voltage component, V_o, into the input of a second BJT amplifier. As will be shown in Chapter 7, we can model the second BJT amplifier (or any amplifier by its equivalent input resistance, R_{in}, to show simply how the second amplifier acts as a load.

Regardless of what the load actually is, we will model it with a resistor and call it R_L. As shown in Fig. 6-7, R_L represents the actual load across which the ac output voltage will be developed. If R_{in} is the actual load, we will call it R_L when analyzing its effect on the original amplifier.

The resistor that is connected from V_{CC} to the BJT's collector no longer

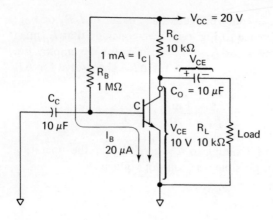

Figure 6-7 Addition of an ac load R_L to the BJT amplifier.

represents the load. Therefore, its name is changed to *coupling resistor*, R_C. We will need R_C to establish an operating point V_{CE} so that V_{CE} can be varied by V_{cep} to develop an output voltage. Output coupling capacitor C_o is required to block the dc component of V_{CE} from appearing across R_L. The role of C_o is explored further in the next section.

6-6.2 Output Coupling Capacitor. C_o in Fig. 6-7 is the output coupling capacitor. Its job is to couple the ac component of v_{CE} as V_o across R_L and block the dc bias component V_{CE}. C_o is selected to be large enough (usually electrolytic) so that it presents negligible reactance to the flow of signal currents. For ac purposes we can think of C_o as a short circuit. Thus any ac voltage developed across R_C will also be developed across R_L.

Although C_o blocks the flow of dc current, we note in Fig. 6-7 that the $(+)$ side of C_o is at a potential of V_{CE}. The $(-)$ side of C_o is at ground potential. This means that C_o must be charged to a potential equal to V_{CE}, or 10 V. The charge on C_o occurs when power is first applied to the circuit. We draw the important conclusion that C_o prevents R_L from having any affect on the dc-operating-point currents of the BJT. That is, if $V_{CE} = 10$ V and $I_C = 1$ mA *before* C_o is connected, then V_{CE} will *still* equal 10 V and $I_C = 1$ mA *after* C_o is connected. Furthermore, since no dc current flows through R_L, the dc potential across R_L must be zero.

6-6.3 AC Load Current. The ac load *on the transistor* of Fig. 6-7 is neither R_L nor R_C but the parallel combination of both. The ac load *on the transistor* is symbolized by R_{ac} and is expressed mathematically by

$$R_{ac} = R_L \| R_C = \frac{R_L \times R_C}{R_L + R_C} \tag{6-5}$$

where $\|$ means "in parallel with."

To understand this principle it is easier to work with collector currents.

First, as shown in Fig. 6-7, the dc collector current I_C flows only through R_C and the BJT. So R_C is the BJT's dc load and determines the dc load line. This dc load is symbolized by R_{dc} to differentiate it from R_{ac}. Mathematically,

$$R_{dc} = R_C \qquad (6\text{-}6)$$

Recall from Section 6-6.2 that C_o did not change the dc operating point, so C_o and R_L do *not* change the dc load line.

Next we know that an ac signal base current will cause an amplified ac signal current, I_c, to flow in the collector lead. But C_o and V_{CC} are short circuits to the flow of ac currents, as shown in Fig. 6-8. This means that I_c will flow through the parallel combination of R_L and R_C. So the total output voltage, V_o, developed across R_{ac} is now

$$V_o = I_c R_{ac} \qquad (6\text{-}7a)$$

or, using peak values,

$$V_{op} = I_{cp} R_{ac} \qquad (6\text{-}7b)$$

Note that I_c divides between R_C and R_L according to Kirchhoff's current law. Rather than using the current division to evaluate I_{RC} and I_{RL} it is easier to evaluate V_o from Eq. (6-7) and then (if necessary) find the load ac current I_{RL} from

$$I_{RL} = \frac{V_o}{R_L} \qquad (6\text{-}8a)$$

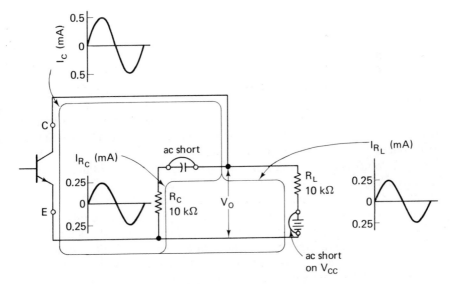

Figure 6-8 Path of ac collector-current components for Fig. 6-7.

and ac coupling resistor current I_{RC} from

$$I_{RC} = \frac{V_o}{R_C} \tag{6-8b}$$

Compare Eq. (6-2b), $V_{op} = I_{cp}R_L$, with Eq. (6-7b) to examine the difference in output voltage for a dc and an ac load, respectively. It is more informative, however, to examine the differences by graphical techniques, which we shall now do.

6-6.4 AC Load Line.
Like the dc load line, the ac load line is plotted on the BJT's output characteristic curves. It also shows how BJT signal voltage V_{ce} varies with signal currents I_b and I_c.

To plot the ac load line we must locate the dc operating point O by first plotting the dc load line. As discussed in Section 6-3, the dc load line is drawn in Fig. 6-9(a) for dc load resistance $R_C = R_{dc}$ in Fig. 6-9(b). We reason that since signal collector current must go through zero twice each cycle, only bias current or operating-point current will flow at these instants. Therefore, the operating point must lie on the ac load line. Next, we reason that any peak ac collector current I_{cp} will produce an ac voltage, V_{op}, across R_{ac} equal to $V_{op} = I_{cp}R_{ac}$ [see Eq. (6-7)]. But in Fig. 6-8, V_o is seen to exist also between the collector and the emitter of the transistor. That is, $V_{op} = V_{cep}$. So we can rewrite Eq. (6-7) as

$$V_{ce} = I_c R_{ac} \tag{6-9a}$$

$$V_{cep} = I_{cp} R_{ac} \tag{6-9b}$$

Equation (6-9b) tells how to plot the ac load line. *Assume* a convenient value for I_{cp} by inspection of the graph in Fig. 6-9(a). Usually $I_{cp} = I_C$ is a good value, so pick $I_{cp} = 1$ mA. Then solve for $R_{ac} = 10$ kΩ//10 kΩ = 5 kΩ from Eq. (6-5). Find the corresponding V_{cep} from Eq. (6-9b) as $V_{cep} = I_{cp}R_{ac} = 1$ mA $\times$ 5 kΩ = 5 V. Starting from operating point O in Fig. 6-9(a) drop a vertical construction line of length equal to I_{cp} to point X. Horizontally, to the right of point X, mark off a distance of $V_{cp} = 5$ V to point Y. Points O and Y are connected and extended as a solid line to locate the ac load line.

Now study Fig. 6-3 to see that a peak base-current signal of $I_{bp} = 10$ μA results (point A) in a peak output voltage $V_{op} = 5$ V. Compare this with the identical dc load (R_L of Fig. 6-3 = R_C of Fig. 6-9) of Fig. 6-9. Note that in both circuits the *same* peak base current causes the *same* peak collector current. Thus in Fig. 6-9(a) the operating point travels along the *ac operating path* from O to A'. But since the ac load is smaller, a smaller voltage change occurs across the transistor. This is seen by dropping construction lines from

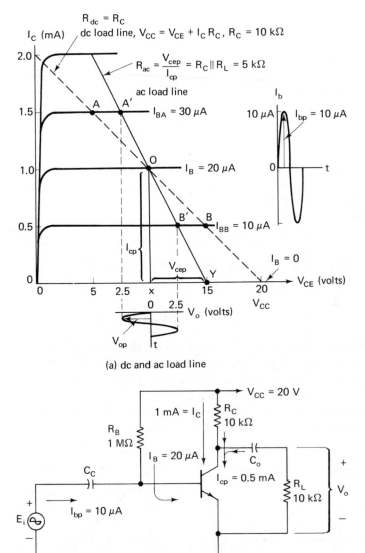

(a) dc and ac load line

(b) Circuit for (a)

Figure 6-9 Ac and dc load lines are drawn in (a) for the circuit of (b). $A'OB'$ locates the ac operating path.

points A' and B'. The peak value of V_{op} is now only 2.5 V. V_{op} was cut in half because the ac load is half the dc load.

6-6.5 Maximum Output Voltage. From review of the principles set forth in Section 6-5 and by reference to Fig. 6-9(a), the maximum output voltage is determined by location of the operating point. However, the conclusions drawn in Section 6-5 must be modified to hold for ac loads as follows: *The maximum allowable peak output voltage is equal to the smaller of V_{CE} or $I_C R_{ac}$,* where V_{CE} and I_C are dc operating-point coordinates. It can be shown that maximum possible peak output voltage occurs if operating-point current is set at

$$I_C = \frac{V_{CC}}{R_{dc} + R_{ac}} \qquad \text{for maximum possible } V_{op} \qquad (6\text{-}10)$$

Example 6-6: What is the (a) maximum available peak output voltage for the circuit of Fig. 6-9, (b) corresponding peak ac load current, and (c) peak ac collector-current component?
Solution:

(a) Since $V_{CE} = 10$ V and $I_C R_{ac} = 1 \times 10^{-3} \times 5 \times 10^3 = 5$ V, $V_{op} = 5$ V.

(b) From Eq. (6-8a), $I_{RL} = V_{op}/R_L = 5$ V/10 kΩ = 0.5 mA.

(c) From Eq. (6-7a), $I_{cp} = V_{op}/R_{ac} = 5$ V/5 kΩ = 1.0 ma.

6-7 Operating-Point Movement

In Sections 5-5 through 5-7 it was shown that biasing a circuit with constant base current had a serious disadvantage. Collector current changed if the transistor β changed. β can change if we substitute transistors. The value of β can also change with temperature variation. Heating a transistor will increase β; cooling it will decrease β. As shown in Fig. 6-10, operating point O moves up the dc load line as β is increased.

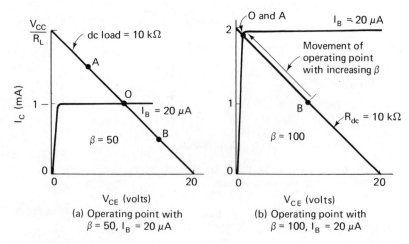

(a) Operating point with
$\beta = 50$, $I_B = 20\ \mu$A

(b) Operating point with
$\beta = 100$, $I_B = 20\ \mu$A

Figure 6-10 Effect of β change on operating point with constant base-current biasing.

A peak base signal current of 10 μA in Fig. 6-10(a) would cause a satisfactory operating path AOB. The same signal in Fig. 6-10(b) would collapse the AO operating path into a single point O. Thus the positive-going base-current signal would cause no collector-current signal, and severe distortion would result. Imagine a radio, with constant-base-bias current, playing in the sun, the operating points moving up their load lines, the music distorting, and then stopping as point B gets to saturation. Remove the radio from the sun and as it cools music once again begins, initially distorted, as the operating points move back down the load line.

6-8 Stabilizing the Operating Point

6-8.1 Stabilizing Action. To stabilize the operating point is to select a bias circuit that minimizes motion of the operating point. The operating point will be stabilized if collector bias current I_C is stabilized. The most common method of stabilizing I_C is to add resistance in series with the emitter terminal. As shown in Fig. 6-11, the emitter resistor is called R_E. When the circuit is properly designed, dc voltage across R_E is about five or more times greater than the $I_B R_B$ voltage drop across R_B. Voltage V_E, across R_E, is approximately equal to $I_C R_E$, since I_E approximately equals I_C. V_E is also the dc emitter voltage with respect to ground.

Stabilizing action occurs in the following way. Suppose that I_C tends to increase (for any reason). Then voltage V_E will tend to increase, raising its opposition to V_{BB} in the base loop. This action reduces I_B. Since I_B is reduced, collector current I_C [which is the product of β and I_B from Eq. (4-2)] tends to reduce, offsetting the increase. This stabilizing action is summarized by saying that base current is automatically adjusted to minimize changes in I_C. As I_C goes up, I_B goes down and vice versa.

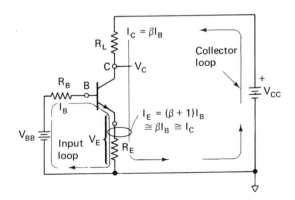

Figure 6-11 Stabilizing I_C with emitter resistor R_E.

6-8.2 Mathematics of Stabilizing Action. As stated in Section 6-8.1 and shown in Fig. 6-11, V_E should be about five times greater than $I_B R_B$. We explore this statement further by writing the input-loop equation,

$$V_{BB} = I_B R_B + V_{BE} + V_E \tag{6-11}$$

In Eq. (6-11a), V_{BE} will equal 0.6 V, and nothing can be done to change it. As will be shown in Chapter 7, R_B is part of the BJT circuit's ac input resistance. Thus R_B should be large to increase input resistance and minimize the signal current which it can draw from a signal generator. But in order for voltage drop $I_B R_B$ to be smaller than V_E, R_B should be small. This is our first encounter with a *tradeoff*. R_B should be small for better stability, yet large for better (higher) input resistance. (The effect of R_B on input resistance is shown in Section 7-7.)

The tradeoff idea may be upsetting when it is first encountered. However, if we are to face reality, it should be clearly understood that improvement of one aspect of circuit performance must always be paid for by downgrading other aspect(s) or by selecting a more sophisticated and expensive circuit. In this tradeoff we can buy stability at the expense of input resistance.

The best way to investigate the tradeoff mathematically is to first substitute in Eq. (6-11) for V_E,

$$V_E = I_C R_E = \beta I_B R_B \tag{6-12}$$

and then solve for I_B:

$$I_B = \frac{V_{BB} - V_{BE}}{R_B + \beta R_E} \tag{6-13}$$

If both sides of Eq. (6-13) are multiplied by β, the left side is βI_B or I_C, and we rearrange the right side to get

$$I_C = \beta I_B = \frac{V_{BB} - V_{BE}}{R_E + (R_B/\beta)} \tag{6-14}$$

Equation (6-14) is the key to understanding how I_C tends to remain constant as β changes. If the term R_B/β is small with respect to R_E, changes in β will not affect I_C. This can be accomplished if R_B *has values that range between* $10R_E$ *and* $20R_E$. These complex ideas will be clarified in the next two examples.

Example 6-7: In Fig. 6-11, $V_{BB} = 0.840$ V, $V_{BE} = 0.6$ V, $R_B = 2000\ \Omega$, and $R_E = 200\ \Omega$. That is, $R_B = 10R_E$. (a) Assume that $\beta = 50$ and find I_C. (b) Assume that a new transistor is installed with $\beta = 100$. Does I_C double as it did with constant-base-current biasing?

Solution:

(a) From Eq. (6-14),

$$I_C = \frac{0.840 - 0.600}{200 + (2000/50)} = \frac{0.240}{200 + 40} = 1 \text{ mA}$$

Note that $R_B/\beta = 40 \ \Omega$ is much smaller than $R_E = 200 \ \Omega$.
(b) From Eq. (6-14),

$$I_C = \frac{0.240}{200 + (2000/100)} = \frac{0.240}{200 + 20} = 1.1 \text{ mA}$$

I_C did not double but only increased by 10%.

Example 6-8: Find the value of I_B, $I_B R_B$, and V_E in parts (a) and (b) of Example 6-7.
Solution:

(a) From Eq. (4-2), $I_B = I_C/\beta = 1$ mA/50 $= 20 \ \mu$A. $V_E = I_C R_E = 1$ mA $\times$ 200 $\Omega = 0.2$ V. $I_B R_B = 20 \times 10^{-6} \times 2 \times 10^3 = 0.04$ V.
(b) $I_B = 1.1$ mA/100 $= 11 \ \mu$A. $V_E = 1.1$ mA $\times$ 200 $\Omega = 0.22$ V. $I_B R_B = 11 \times 10^{-6} \times 2 \times 10^{-3} = 0.02$ V.

We conclude from Examples 6-7 and 6-8 that I_C changed only 10% for a 100% change in β. I_B was automatically reduced by V_E to hold I_C almost constant. A simple *approximation* rule can be stated that for a 100% change in β, if $R_B = 10R_E$, I_C will change by 10%. If $R_B = 20R_E$, I_C will change by 20% and so on.

6-8.3 Practical Biasing with an Emitter Resistor. In Fig. 6-11, the use of a separate battery V_{BB} for biasing is clumsy and expensive. We cannot simply substitute V_{CC} for V_{BB} because R_B would be much too large for effective stabilizing action. Fortunately, we can use Thévenin's theorem in reverse to find a more practical substitute for R_B and V_{BB}.

The Thévenin circuit of Fig. 6-12(a) is an exact equivalent of the practical version in Fig. 6-12(b) if

$$R_B = R_1 \| R_2 = \frac{R_1 R_2}{R_1 + R_2} \tag{6-15a}$$

$$V_{BB} = V_{CC} \frac{R_2}{R_1 + R_2} \tag{6.15b}$$

where

$$R_1 = \frac{V_{CC}}{V_{BB}} R_B \quad \text{and} \quad R_2 = \frac{V_{CC}}{V_{CC} - V_{BB}} R_B$$

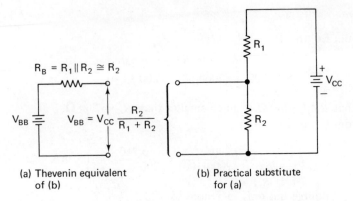

(a) Thevenin equivalent
of (b)

(b) Practical substitute
for (a)

Figure 6-12 The base-bias circuit of Fig. 6-11 in (a) may be replaced by the more practical arrangement in (b).

These equations are useful if values for β, R_1, and R_2 are known precisely together with R_L, R_E, and V_{CC}. Usually standard resistors are used and have an inherent 5 or 10% variation in their standard value. Also the standard resistors will probably not be available in sizes required for R_1 and R_2 by these equations. Furthermore, as shown in Section 6-8.2, the voltage drop across R_B is small with respect to emitter voltage V_E. Hence V_{BB} is *approximately* equal to base voltage V_B, where V_B is the dc voltage from base to ground. Also V_B will usually be small with respect to V_{CC}. Using these approximations simplifies the four equations of (6-15) to

$$R_2 \cong R_B \tag{6-16a}$$

$$R_1 = \frac{V_{CC}}{V_B} R_B \tag{6-16b}$$

$$V_B \cong V_{CC} \frac{R_2}{R_1 + R_2} \tag{6-16c}$$

These approximate equations are shown in the final practical circuit of Fig. 6-13, and their application is demonstrated in the next examples.

Example 6-9: In Fig. 6-13, $R_E = 200\ \Omega$, $R_L = 10\ \text{k}\Omega$, and $V_{CC} = 20\ \text{V}$. Choose R_1 and R_2 to stabilize I_C at 1 mA by making R_B approximately equal to $10R_E$.
Solution:

(a) Pick a standard resistor for R_B of 2200 Ω or ($R_B = 11R_E$). From Eq. (6-16a) pick $R_B = R_2 = 2200\ \Omega$.

(b) Since we do not know R_1, we cannot use Eq. (6-16c). But from Fig. 6-13, V_B supplies both V_{BE} and V_E, or

$$V_B = V_{BE} + V_E \tag{6-17}$$

Solve for V_E from $V_E = I_C R_E = 1 \text{ mA} \times 200\ \Omega = 0.2$ V and substitute into Eq. (6-17),

$$V_B = 0.6 + 0.2 = 0.8 \text{ V}$$

(c) From Eq. (6-16b),

$$R_1 = \frac{V_{CC}}{V_B} R_B = \frac{20}{0.8} \times 2200 = 55 \text{ k}\Omega$$

Choose the nearest standard-sized resistor of $R_1 = 56$ kΩ.

Example 6-10: In Fig. 6-13, $R_L = 10$ kΩ, $R_E = 200\ \Omega$, $V_{CC} = 20$ V, $R_1 = 100$ kΩ, and $R_2 = 3.9$ kΩ. Find the approximate value for I_C.
Solution: From Eq. (6-16c),

$$V_B = 20 \frac{3.9 \times 10^3}{(3.9 + 100) \times 10^3} = \frac{20 \times 3.9}{104} = 0.75 \text{ V}$$

Find V_E from Eq. (6-17):

$$V_E = V_B - V_{BE} = 0.75 - 0.60 = 0.15 \text{ V}$$

Finally, $I_C = V_E / R_E = 0.15/200 = 0.7$ mA.

6-8.4 Measuring and Refining the Operating Point. Do not conclude from the design Example, 6-9, and the analysis Example, 6-10, that the results of our

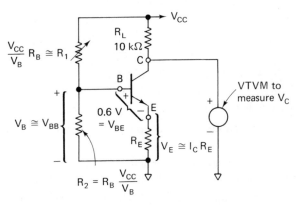

Figure 6-13 Practical bias circuit to stabilize I_C.

calculations will give a circuit that delivers the specified collector current. If R_2 is a 10% resistor, its value could range from 2 to 2.4 kΩ in Example 6-9 and from 3.5 to 4.3 kΩ in Example 6-10. The same problem can exist for the other resistors, and V_{BE} might be 0.55 V or even 0.65 V. So when the circuit is built, its operating point should be measured.

The best single check of the operating point is to measure the dc voltage, V_C, from collector to ground. This allows use of a VTVM that normally has one terminal grounded, as in Fig. 6-13. Interpretation of the reading for V_C is shown by an example.

Example 6-11: In Example 6-9 the required value for I_C was 1 mA. In the actual circuit, V_C is measured as 15 V. (a) Is I_C too high or two low? (b) What can be done to R_1 or R_2 to refine I_C back to 1 mA?
Solution:

(a) If I_C were 1 mA, voltage across R_L would be 10 V and V_C would be $V_{CC} - V_{RL} = V_C = 10$ V. But V_{RL} actually equals $V_{CC} - V_C = 20 - 15 = 5$ V. So I_C actually equals $V_{RL}/R_L = 5$ V/10 kΩ = 0.5 mA. I_C is too low.
(b) If I_C is too low, V_E is too low, so V_B must also be too low. To raise V_B we must increase R_2 slightly or decrease R_1 slightly so that more of V_{CC} is developed across R_2. In practice it is usually simpler to decrease a resistor by connecting (trial and error) a much larger resistor across it.

We conclude that if V_C is too high, increase R_2; and if V_C is too low, decrease R_2. The opposite action can be taken with R_1.

Problems

6-1 What is the purpose of C_c in Fig. 6-2?

6-2 Plot the instantaneous base-current wave form if $I_B = 30$ μA and $I_{bp} = 10$ μA.

6-3 For the values given in Problem 6-2, what is the (a) maximum value of base current; (b) minimum value of base current?

6-4 Use Eq. (6-2b) to calculate V_{op} for Fig. 6-2.

6-5 The circuit of Fig. 6-4 has a gain of 333 (calculated from Example 6-1). If $E_{ip} = 5$ mV, what is the peak-to-peak output voltage?

6-6 What would be the value of output voltage read on an ac VTVM for Problem 6-5?

6-7 Input voltage to a common-emitter amplifier is 10 mV, and the output voltage is 1.5 V. Calculate the gain (both values are rms values).

6-8 Refer to Fig. 6-5 and calculate β_{4c} if the operating point is at (a) point A; (b) point B.

6-9 From the results of Problem 6-8, can β_{dc} change if the operating point is changed?

6-10 Refer to Fig. 6-3 and calculate (a) β_{ac} and (b) β_{dc} at the operating point O.

6-11 If a transistor's base-current steps are equally spaced, will $\beta_{dc} = \beta_{ac}$?

6-12 The dc base current, I_B, equals 25 μA. If the transistor has $\beta = 60$, calculate the collector current, I_C.

6-13 What is the typical range of saturation voltage for a BJT?

6-14 What is the advantage for biasing a transistor with the operating point in the center of the load line?

6-15 If the operating point in Example 6-4 is at $I_C = 2.5$ mA and $V_{CE} = 9$ V, find (a) maximum available peak output voltage V_{op}; (b) peak ac collector current I_{cp}; (c) peak ac base current I_{bp}.

6-16 In Fig. 6-7, if $R_c = 10$ kΩ and $R_L = 5$ kΩ, what is the actual ac load on the transistor?

6-17 If R_L in Fig. 6-7 is changed (such as to 5 kΩ), does the dc voltage across C_o change? (R_c remains equal to 10 kΩ.)

6-18 For the circuit of Fig. 6-7, let $R_c = 10$ kΩ, $R_L = 5$ kΩ, and peak ac collector current $I_{cp} = 0.5$ mA. Calculate peak output voltage V_{op}.

6-19 If $R_L = 10$ kΩ and $R_c = 5$ kΩ in Fig. 6-7 and $V_o = 2$ V, what is (a) ac current through R_c; (b) ac current through R_L?

6-20 What two variations in a transistor amplifier may cause changes in the beta of the transistor?

6-21 Is the dc collector current stabilized (does not change) if I_B is held constant?

6-22 Refer to Fig. 6-11 and let $V_{BB} = 1.175$ V, $R_B = 1.5$ kΩ, $R_E = 100$ Ω, and $V_{BE} = 0.6$ V. Calculate (a) dc base current, I_B; (b) dc collector current, I_C. $\beta = 100$.

6-23 In Example 6-7, if $\beta = 200$, calculate dc collector current, I_C.

6-24 Choose R_1 and R_2 in Example 6-9 to stabilize I_c at 2 mA.

6-25 If R_2 in Example 6-10 is changed to 5.6 kΩ, calculate dc collector current, I_C.

7

Common-Emitter
Small-Signal Operation

7-0 Introduction

Voltage gains and input resistance are the keys to understanding principles of transistor operation. In the beginning there is a signal voltage E_i. When it is applied to a circuit, the circuit can be modeled by an equivalent resistance, R_{in}, that it presents to the signal voltage. If it is necessary to find the current, I_{in}, drawn from E_i, we simply apply Ohm's law, $I_{in} = E_i/R_{in}$.

If E_i is amplified by a circuit that gives a voltage gain A_v, an output voltage V_o is developed across a load R_L. V_o will equal $A_v E_i$. So if we know the voltage gain, we can calculate V_o. And if we know V_o, we can find load current with Ohm's law.

As observed in Section 6-1, the graphical technique is not suitable for determining input resistance. Also, while a graphical technique is a good visual aid to explain ideas of voltage gain and how the ac signal currents ride on the dc bias currents, it is slow and tedious. Fortunately, there is an analysis technique that allows us, with *reasonable precision*, to *estimate* voltage gains and input resistances without using elaborate charts and graphs. To introduce this technique we begin by amplifying the idea of modeling.

7-1 Small-Signal BJT Model

7-1.1 Requirements of the Model. Refer to the CE circuit of Fig. 7-1. Dc bias currents and voltages can be found by methods given in Chapters 5 and 6. E_i is a known ac signal voltage. How can we find the values of V_o or, even

108

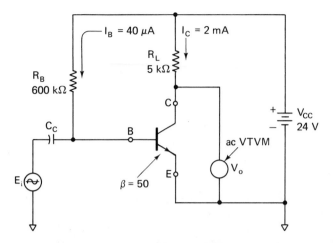

Figure 7-1 Common-emitter amplifier. $V_o = A_v E_i$.

better, voltage gain $A_v = V_o/E_i$? We can use a model of the BJT that explains how it behaves with ac currents and voltages. Since dc circuit behavior is already known, the BJT model need *only* be valid for ac circuits and voltages.

The model should allow us to evaluate how much ac current is drawn by the BJT's input terminal from E_i. It would be even better if the model could give us a value for the *transistor's input resistance*, R_i. Then we could evaluate input current for any value of E_i, by use of Ohm's law. The model should also tell what ac output current results from the ac input current. Finally, with a little ingenuity we should be able to use the model to get a simple equation that gives voltage gain A_v without the need to evaluate input and output currents.

7-1.2 Layout of the Model. Only two circuit elements make up the small-signal model in Fig. 7-2b. This model represents the ac behavior of both *npn* and *pnp* BJTs. It also makes no difference if the BJTs are made of silicon or germanium.

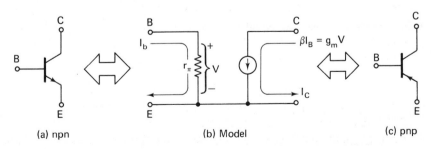

(a) npn   (b) Model   (c) pnp

Figure 7-2 Small-signal model in (b) for either *npn* or *pnp* BJTs.

When an ac signal voltage, V, is developed between base and emitter terminals, it sees some resistance to the ac input current I_b that flows between base and emitter terminals. This resistance is modeled by r_π in Fig. 7-2(b) and is expressed as

$$r_\pi = \frac{V}{I_b} \qquad (7\text{-}1)$$

An output ac signal current I_c flows between collector and emitter. I_c *depends* on I_b according to Eq. (6-4d) as

$$I_c = \beta I_b \qquad (6\text{-}4d)$$

That is, when I_b increases, I_c increases, because β is constant. So βI_b is modeled by a *dependent* current generator symbol in Fig. 7-2.

7-1.3 Transconductance of a BJT. An important electrical characteristic of the BJT (as well as of vacuum tubes and field-effect transistors) is its *transconductance*, g_m. "Trans" signifies a relationship between output and input. "Conductance" means a relationship between current and voltage. Putting these fundamentals together gives the definition of transconductance for a BJT. Transconductance, g_m, is the ratio of ac output collector current I_c to ac input voltage V, or $g_m = I_c/V$. To show how transconductance acts in the model of Fig. 7-2(b), we rewrite the definition as

$$I_c = g_m V \qquad (7\text{-}2a)$$

And, since I_c also equals βI_b, we equate the expressions for I_c as

$$\beta I_b = g_m V \qquad (7\text{-}2b)$$

There is a simple relationship among g_m, r_π, and β. Compare Eq. (7-1) with a revised form of Eq. (7-2b):

$$(7\text{-}1) \quad r_\pi = \frac{V}{I_b} \qquad (7\text{-}2b) \quad \frac{V}{I_b} = \frac{\beta}{g_m}$$

We conclude that they are equal and, therefore,

$$r_\pi = \frac{\beta}{g_m} \qquad (7\text{-}3)$$

7-2 Evaluating the Small-Signal Model

7-2.1 Evaluation of Beta. There are only two electrical characteristics that must be evaluated in the small-signal model, β and r_π. The *only* BJT electrical characteristic that must be obtained by either measurement or from data sheets is β. Technically we should use the ac beta, or h_{fe}. However, as stated in Section 6-4, we will use the dc beta (from the operating point) as given by Eq. (4-2) where $\beta = I_C/I_B$. Note that a value for β must have already have been obtained when analyzing the dc bias problem. Therefore, we already know half of the ac small-signal model when the dc portion of the problem is completed.

7-2.2 Evaluation of Transconductance. To evaluate transconductance, g_m, all we need to know is the value of the dc operating-point collector current I_C. At room temperatures g_m is found approximately, but simply, from

$$g_m = \frac{I_C \text{ (amperes)}}{0.030 \text{ V}} \tag{7-4a}$$

or

$$g_m = \frac{I_C \text{ (mA)}}{30 \text{ mV}} \tag{7-4b}$$

Equation (7-4) is valid for both silicon and germanium transistors. We will eventually use g_m to find r_π or voltage gain. Therefore, it is recommended that g_m *not* be evaluated in decimal form but be left in fraction form, as shown in the following example.

Example 7-1: Evaluate g_m for circuits with collector bias currents of (a) $I_C = 0.1$ mA, (b) $I_C = 1$ mA, and (c) $I_C = 10$ mA.
Solution: From Eq. (7-4b),
(a) $g_m = \dfrac{0.1 \text{ mA}}{30 \text{ mV}} = \dfrac{1}{300} \mho$
(b) $g_m = \frac{1}{30} \mho$
(c) $g_m = \frac{10}{30} = \frac{1}{3} \mho$

Note that the units for g_m are mhos ($\mho$).

7-2.3 Evaluation of r_π. Equation (7-3) is used to evaluate r_π (after g_m is evaluated from Section 7-2.2). The procedure is illustrated by an example.

Example 7-2: Find r_π for the BJT in the circuit of Fig. 7-1, where $\beta = 50$ and $I_C = 2$ mA.
Solution:
(a) Eq. (7-4b), $g_m = \frac{2}{30}$ mho.
(b) Eq. (7-3),

$$r_\pi = \frac{\beta}{g_m} = \frac{50}{\frac{2}{30} \mho} = \frac{50 \times 30}{2 \mho} = \frac{25 \times 30}{\mho} = 750 \ \Omega$$

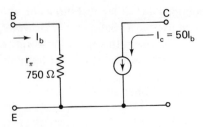

Figure 7-3 Small-signal model for the BJT in Fig. 7-1 and Example 7-2.

Observe how the fraction g_m is handled in part (b) to reduce the chance of calculating errors, and how the units for r_π are in ohms. The resulting model is shown in Fig. 7-3.

7-2.4 Alternative Evaluation of r_π. In many applications we do not need to evaluate g_m. It is more direct to find r_π directly from I_C by combining Eqs. (7-3) and (7-4b) as

$$r_\pi = \frac{30\beta}{I_C \text{ (mA)}} \qquad (7\text{-}5\text{a})$$

For example, if $I_C = 2$ mA and $\beta = 50$,

$$r_\pi = 30 \times \frac{50}{2} = 750 \ \Omega$$

By substituting I_B for I_C in Eq. (7-5a), we learn how r_π can be found directly from the dc base-bias current I_B, where

$$r_\pi = \frac{30}{I_B \text{ (mA)}} \qquad (7\text{-}5\text{b})$$

In this example, if $I_B = I_C/\beta = 2 \text{ mA}/50 = 40 \ \mu\text{A}$, or $I_B = 40 \ \mu\text{A} = 0.040$ mA, then

$$r_\pi = \frac{30}{0.040} = \frac{30 \times 1000}{40} = 750 \ \Omega$$

Equation (7-5a) is often more useful than either Eq. (7-5b) or (7-3), because we always know the value of dc collector bias current, I_C.

7-3 The BJT and Circuit Model

Now that we know what the small-signal model does (Section 7-1) and how to evaluate it (Section 7-2), it is time to use it. Recall that we use this model to study ac behavior of a BJT in a circuit. Thus we replace the BJT symbol in the schematic with its small-signal model. It is equally important that we replace the circuit with its equivalent ac model. For simplicity we assume, and perform tests with, an ac signal frequency of 1000 Hz. This means that (1) *coupling capacitors can be modeled by a short circuit,* and (2) *supply voltages are replaced by their ac impedance at 1000 Hz, that is, usually a short circuit.*

We use Figs. 7-1 and 7-3 to illustrate the procedure. First, refer to the circuit schematic redrawn in Fig. 7-4(a). Next draw the BJT model in Fig. 7-4(b). Then replace the battery symbol for V_{CC} by a short circuit between

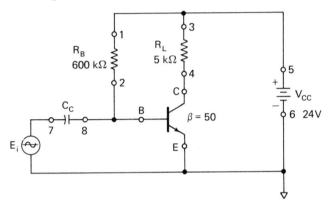

(a) Common emitter circuit

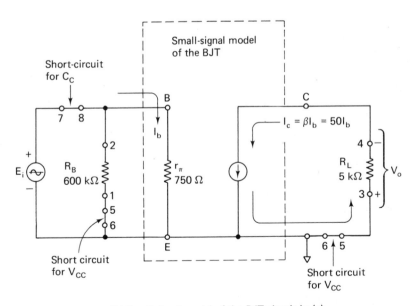

(b) Small-signal model of the BJT circuit in (a)

Figure 7-4 The BJT circuit in (a) is modeled for ac signal analysis in (b).

terminals 5 and 6. Also replace the symbol for C_C by a short circuit between terminals 7 and 8.

Figure 7-4(b) should be studied with care to emphasize the following facts:

1. Terminals 1 of R_B and 3 of R_L are placed at ac ground by the zero impedance of V_{CC}.
2. Terminal 2 of R_B and the BJT's base terminal are connected directly to E_i by the zero impedance of C_C.

From these facts we draw the following conclusions:

1. Input voltage E_i and r_π determine how much base signal current, I_b, is drawn by the BJT. R_B does *not* affect I_b. R_B merely wastes signal current from E_i. Remember that R_B is an evil that we must have to set the dc bias currents. So we conclude that R_B should be high to minimize loading on E_i.
2. Ac output voltage V_o is developed across load R_L, and one side of V_o is at ac ground.

This completes all the material necessary for the development of the ideas of voltage gain and input resistance.

7-4 Input Resistance of the Common-Emitter Circuit

Once we have learned how to draw the ac circuit model and BJT model together, as in Fig. 7-4(b), it is natural to omit the terminal numbers and present the simplified version of Fig. 7-5(a). There are two input resistances of interest. The first is the resistance presented to E_i by the *transistor*. It is identified in Fig. 7-5(a) by the symbol R_i. The second is the resistance presented to E_i by the *circuit*. It is identified by the symbol R_{in} and, as shown in Fig. 7-5(b), is equal to the parallel combination of R_B and R_i. Mathematically,

$$R_i = r_\pi = \frac{E_i}{I_b} \tag{7-6}$$

$$R_{in} = R_B \| R_i = \frac{E_i}{I_i} \tag{7-7}$$

Example 7-3: For Fig. 7-5 calculate the resistances presented to E_i by the (a) transistor and (b) circuit.
Solution:

(a) From Eq. (7-6), $R_i = r_\pi = 750 \ \Omega$.

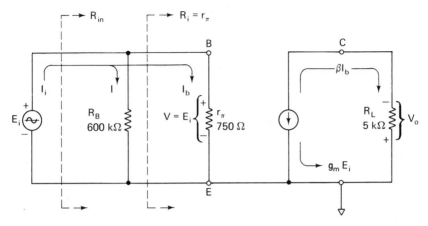

(a) Small-signal model of Fig. 7-4(a)

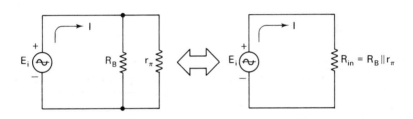

(b) Input resistance of circuit

Figure 7-5 Input resistance and voltage gain of a CE circuit. $\beta = 50$, $g_m = \frac{2}{30} \, \mho$.

(b) From Eq. (7-7),

$$R_{\text{in}} = R_B \| R_i = \frac{R_B \times R_i}{R_B + R_i} = \frac{600{,}000 \times 750}{600{,}750} \simeq 750 \, \Omega$$

(*Note:* Because R_B in this circuit is so large, R_{in} is approximately R_i. However, for other types of biasing arrangements, R_B may not be so large, and R_{in} will *not* equal R_i.)

7-5 Voltage Gain of the Common-Emitter Circuit

7-5.1 Introduction. It would be helpful to have one gain equation that would tell almost everything about the circuit and include the input resistance, current gain, and voltage gain. Suppose there were also a pattern by which we could construct this most useful equation by merely *looking at the circuit*. We would then have a most powerful tool that would not only free us from memorizing lists of equations but would also allow us to visualize how the circuit works.

7-5.2 Standard Form of Voltage Gain. Fortunately, there is such a pattern. To get at it we first dispose of R_B. In Fig. 7-5a R_B has no effect on I_b. Therefore, R_B has no effect on I_b and cannot affect V_o. We conclude that R_B has no effect on gain and should not appear in the voltage-gain equation. Next we make two observations. First, V_o is the product of collector signal current βI_b and R_L. Second, E_i equals the product of R_i and I_b. Arrange these two equations as follows:

$$V_o = \beta I_b R_L \qquad (7\text{-}8)$$
$$E_i = I_b R_i \qquad (7\text{-}6a)$$

Dividing these equations and canceling I_b gives the standard form:

$$A_v = \frac{V_o}{E_i} = \beta \frac{R_L}{R_i} \qquad (7\text{-}9)$$

where *voltage gain* A_v is the ratio of output voltage V_o to input voltage E_i.

There is a definite pattern to Eq. (7-9). The β term was obtained from the ratio of BJT *output* current βI_b to BJT input current I_b. Therefore, the term β represents *current gain* of the BJT. The expression for current gain is easily constructed. In goes I_b and out comes βI_b, so their ratio is current gain β. R_L is the ac load resistance across which V_o is measured. R_i is the resistance *of the transistor*. Both R_i and R_L are easily identified in the circuit model. So the pattern in Eq. (7-9) is

$$A_v = \frac{V_o}{E_i} = \text{current gain} \times \frac{\text{ac load resistance}}{\text{BJT input resistance } R_i} \qquad (7\text{-}10)$$

Once the pattern is mastered, the voltage-gain expression can be constructed *directly from inspection of the circuit*.

Example 7-4: Construct the voltage-gain equation directly from the circuit of Fig. 7-5(a).
Solution: By inspection, A_v equals

(1) current gain = (collector current)/(base current) = β; times
(2) ac load resistance, R_L; divided by
(3) BJT input resistance, R_i, that is, equal to r_π:

$$A_v = \frac{V_o}{E_i} = \beta \frac{R_L}{r_\pi} \qquad (7\text{-}11a)$$

Example 7-5: Find another form for Eq. (7-11a) by comparison with Eq. (7-3), and calculate the voltage gain for Fig. 7-5.

Solution: Note from Eq. (7-3) that $g_m = \beta/r_\pi$. Substituting for β/r_π in Eq. (7-11a),

$$A_v = \frac{V_o}{E_i} = g_m R_L \qquad (7\text{-}11b)$$

Evaluating $A_v = \frac{2}{30} \times 5000 = 333$: if E_i was 1 mV, then V_o would equal $A_v E_i = 333 \times 1$ mV $= 333$ mV. Equation (7-11b) is useful in showing how gain A_v depends on operating-point current. As I_C increases, g_m increases, and so does gain.

Example 7-6: Using Eq. (7-11a), calculate voltage gain for the circuit of Fig. 7-5 and compare the results with Example 7-5.

Solution: From Eq. (7-10),

$$A_v = \beta \frac{R_L}{r_\pi} = 50 \times \frac{5000}{750} = 333$$

7-6 Voltage Gain with Capacitor-Coupled Load

In Section 6-6 graphical analysis was used to learn how an ac load changed the performance of the basic amplifier with a dc load. Section 6-6 should be reviewed and the results obtained in Fig. 6-9 compared with the following small-signal analysis.

The circuit of Fig. 6-9(b) is redrawn for convenience in Fig. 7-6(a). We want to evaluate voltage gain and the resistances presented to E_i. This will be accomplished with the aid of examples and serve as a review of the principles developed thus far.

Example 7-7: Draw the small-signal model of the circuit in Fig. 7-6(a) by applying the procedure given in Section 7-3.

Solution:

(a) The battery symbol for V_{CC} is not shown in Fig. 7-6 as it was in Fig. 7-4. We reason that the $(+)$ terminal of V_{CC} must be connected to the arrowhead above R_C in Fig. 7-6(a). The $(-)$ terminal of V_{CC} would be connected to the ground symbol. To draw the ac model we visualize an ac short across V_{CC} or between the arrowhead and ground. This short connects R_{BB} and R_C to the ground in Fig. 7-6(b).

(b) Coupling capacitor C_C connects one terminal of R_C to one terminal of R_L with an ac short. The other terminals of R_L and R_C are grounded. There-

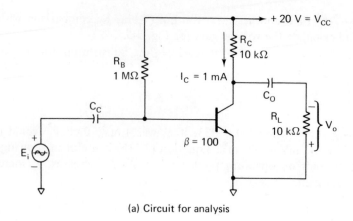

(a) Circuit for analysis

(b) Small-signal model of (a)

(c) R_C and R_L in (b) are replaced by their
equivalent parallel load resistance R_{ac}

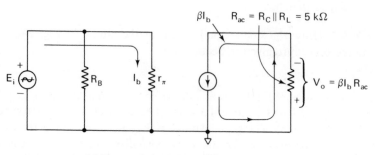

Figure 7-6 The circuit in (a) is modeled in (b) and simplified in (c) for
ac analysis.

fore, in Fig. 7-6(b), coupling resistor R_C and load resistor R_L are connected in parallel.

Example 7-8: How must gain equation (7-9) be modified so that it can be applied to Fig. 7-6(b)?

Solution: In Fig. 7-6(b), V_o is developed by collector signal current flowing through the *parallel combination* of R_C and R_L. Back in Eq. (6-5), this parallel combination was defined as the *ac load* R_{ac} and given by

$$R_{ac} = R_C \| R_L = \text{ac load}$$

So in our standard form Eq. (7-10), R_{ac} is the load that V_o is developed across, and we substitute R_{ac} for R_L in the equation

$$A_v = \beta \frac{R_{ac}}{R_i} \tag{7-9}$$

Example 7-9: For Fig. 7-6, evaluate (a) g_m, (b) r_π, (c) R_i and R_{in}, (d) R_{ac}, and (e) A_v, by two methods.

Solution:

(a) Eq. (7-4b), $g_m = \frac{1}{30}$

(b) Eq. (7-3), $r_\pi = \dfrac{100}{\frac{1}{30}} = 100 \times 30 = 3 \text{ k}\Omega$

(c) $R_i = r_\pi = 3 \text{ k}\Omega$; Eq. (7-6b),

$$R_{in} = \frac{3 \text{ k}\Omega \times 1 \text{ M}\Omega}{3 \text{ k}\Omega + 1 \text{ M}\Omega} \simeq 3 \text{ k}\Omega$$

(d) Eq. (6-5),

$$R_{ac} = \frac{R_c \times R_L}{R_c + R_L} = \frac{(10 \times 10) \text{ k}\Omega}{(10 + 10) \text{ k}\Omega} = 5 \text{ k}\Omega$$

(e) Eq. (7-9),

$$A_v = 100 \times \tfrac{5000}{3000} = 166$$

Eq. (7-11b), $A_v = g_m R_{ac} = \frac{1}{30} \times 5000 = 166$

7-7 Common-Emitter Circuit with an Emitter Resistor

7-7.1 Effect of R_E and Bias Resistors on Input Resistance. Recall from Section 6-8.3 that emitter resistor R_E was added to stabilize dc collector current against changes in β. Also, two biasing resistors, R_1 and R_2, had to be used to make the circuit practical, as in Fig. 7-7(a).

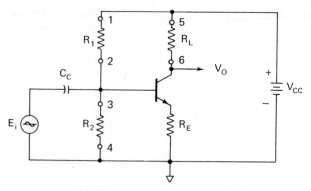

(a) CE circuit with emitter resistance

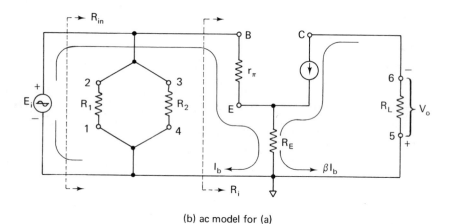

(b) ac model for (a)

Figure 7-7 Circuit and model to study effects of emitter resistance R_E on input resistance and voltage gain.

Input resistance of the transistor, R_i, will be shown to equal

$$R_i = r_\pi + \beta R_E \qquad (7\text{-}12)$$

Input resistance of the circuit, R_{in}, is given by

$$R_{in} = R_B \| R_i \qquad (7\text{-}13)$$

where R_i is found from Eq. (7-12) and R_B is found from Eq. (6-15a):

$$R_B = R_1 \| R_2 \qquad (6\text{-}15a)$$

We conclude from Eq. (7-12) that adding R_E will *increase* input resistance to the transistor. However, as shown in the next example, it is possible for R_B to have a significant effect on R_{in}.

Example 7-10: In Fig. 7-7(a), $R_2 = 2200\ \Omega$, $R_1 = 55\ k\Omega$, $\beta = 50$, $R_L = 10\ k\Omega$, $R_E = 200\ \Omega$, $V_{CC} = 20\ V$, and $I_C = 1\ mA$ (see Example 6-9). Find (a) r_π, (b) R_i, (c) R_B, and (d) R_{in}.
Solution:

(a) From Eq. (7-5a),

$$r_\pi = \frac{30\beta}{I_C} = \frac{30 \times 50}{1} = 1500\ \Omega$$

(b) Eq. (7-12), $R_i = r_\pi + \beta R_E = 1500 + 50 \times 200 = 11.5\ k\Omega$

(c) Eq. (6-15a),

$$R_B = \frac{R_1 R_2}{R_1 + R_2} = \frac{(55 \times 2.2)\ k\Omega}{(55 + 2.2)\ k\Omega} = 2.1\ k\Omega$$

(d) Eq. (7-13),

$$R_{in} = \frac{R_B R_i}{R_B + R_i} = \frac{(2.1 \times 11.5)\ k\Omega}{(2.1 + 11.5)\ k\Omega} = 1.8\ k\Omega$$

Example 7-11: How would the results of Example 7-10 change if β were doubled to 100?

(a) r_π doubles to 3 kΩ.
(b) R_i doubles to $r_\pi + \beta R_E = 3\ k\Omega + 100 \times 200\ \Omega = 23\ k\Omega$.
(c) R_B is unchanged.
(d) R_{in} increases slightly, to $(2.1 \times 23)\ k\Omega/(2.1 + 23)\ k\Omega = 1.9\ k\Omega$.

We learn from Examples 7-10 and 7-11 that R_B (really R_2) controls the *circuit's* input resistance. The principle of resistance multiplication by a BJT that is responsible for Eq. (7-12) will be explained in Section 7-8.

7-7.2 Effect of R_E on Voltage Gain. An excellent approximation for voltage gain in the circuit of Fig. 7-7 is simply

$$A_v = \frac{V_o}{E_i} = \frac{R_L}{R_E} \tag{7-14a}$$

A more exact equation is constructed from the standard form by using Eq. (7-10); I_b goes in and βI_b comes out of the BJT, so the current gain is β. The resistance presented to V_o is R_L. BJT resistance, presented to E_i, is R_i, as given by Eq. (7-12). Thus

$$A_v = \frac{V_o}{E_i} = \beta \frac{R_L}{R_i} = \beta \frac{R_L}{r_\pi + \beta R_E} \tag{7-14b}$$

We note that if r_π is small with respect to βR_E in Eq. (7-14b), r_π can be neglected. The β terms will cancel and Eq. (7-14b) simplifies to Eq. (7-14a).

Example 7-12: In Fig. 7-7, $r_\pi = 1.5\ \text{k}\Omega$, $R_i = 11.5\ \text{k}\Omega$, $\beta = 50$, $R_L = 10\ \text{k}\Omega$, and $R_E = 200\ \Omega$. Find the voltage gain by both approximate and exact methods.
Solution: From Eq. (7-14a), $A_v = 10{,}000/200 = 50$; from Eq. (7-14b), $A_v = 50 \times (10{,}000/11{,}300) = 44$

Make the important observation from Eq. (7-14a) that voltage gain is stabilized by R_E against changes in β. A_v will depend on the stable resistors R_L and R_E. If β is doubled, gain will change very little, as shown by comparing Example 7-12 with the next example.

Example 7-13: The transistor is changed in Example 7-12 to one with $\beta = 100$. R_L and R_E remain the same, but r_π increases to 3 kΩ and R_i to 23 kΩ (from Example 7-11). Does voltage gain double?
Solution: From Eq. (7-14b), $A_v = 100(10{,}000/23{,}000) = 43$. The voltage gain is barely changed, and we conclude that R_E stabilizes the circuit against both β and operating-point changes.

7-7.3 Effect of $\mathbf{R}_E$ on Maximum Output Voltage. Conclusions drawn in Section 6-6.5 on maximum available peak output voltage, max V_{op}, must be modified to include the effects of an unbypassed emitter resistor. For dc loads such as shown in Fig. 7-4, maximum V_{op} is the smaller of

$$\text{max } V_{op} = DV_{CE} \quad \text{or} \quad \text{max } V_{op} = DI_C R_L \qquad (7\text{-}15a)$$

where

$$D = \frac{R_L}{R_L + R_E}$$

For ac loads such as shown in Fig. 7-9, with switch 1 down and switch 2 up, max V_{op} is the smaller of

$$\text{max } V_{op} = D_a V_{CE} \quad \text{or} \quad \text{max } V_{op} = D_a I_C R_{ac} \qquad (7\text{-}15b)$$

where $D_a = R_{ac}/(R_{ac} + R_E)$ and R_{ac} is the parallel combination of coupling and load resistor.

7-8 Resistance Multiplication

Focus attention on the input of Fig. 7-7(b) by redrawing as in Fig. 7-8(a). From Kirchhoff's voltage law around the I_b loop,

$$E_i = I_b r_\pi + I_e R_E$$

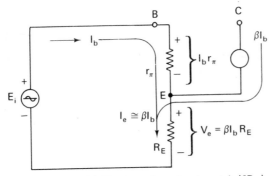

(a) Circuit to show $E_i = I_b r_\pi + I_b(\beta R_E)$

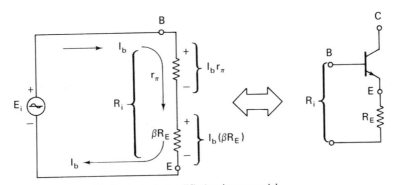

(b) Equivalent simplified ac input model

Figure 7-8 Resistor R_E in the emitter leg of (a) is transformed as βR_E into the base leg of (b).

But since $I_e \cong \beta I_b$, the drop across R_E can be given in terms of I_b only as $(\beta I_b) R_E$, so

$$E_i = I_b r_\pi + (\beta I_b) R_E$$

There is no reason why we cannot rearrange terms again to write the drop across R_E as $I_b(\beta R_E)$. This is no idle exercise, because now we can think of the drop across R_E in terms of signal base current $I_b \times$ multiplied resistance βR_E. That is, I_b *"sees" any resistance in the emitter leg as an apparently larger resistance* $\beta \times R_E$ *in the base leg.*

$$E_i = I_b r_\pi + I_b(\beta R_E) = I_b(r_\pi + \beta R_E)$$

Rearranging terms gives

$$\frac{E_i}{I_b} = R_i \cong r_\pi + \beta R_E \tag{7-12}$$

This is the principle of *resistance multiplication* by a BJT. Put a resistor R_E in the emitter leg and the BJT makes R_E appear much larger (as βR_E) when we look into the base terminal. The multiplication principle always applies and is stated as a rule: *When looking into the base terminal of a BJT, any emitter-leg resistance is multiplied by β.* Thus, as shown in Fig. 7-8(b), an equivalent input resistance can be drawn immediately and directly from an inspection of the circuit. Note that only I_b flows in this equivalent input circuit.

Example 7-14: A BJT has $\beta = 100$ and $r_\pi = 3\ k\Omega$ in a circuit with $R_E = 200\ \Omega$. What resistance does the BJT present to E_i if E_i is connected to the base terminal?

Solution: R_E will be transformed into the base leg as $\beta R_E = 100 \times 200\ \Omega = 20\ k\Omega$ and will be in series with $r_\pi = 3\ k\Omega$. As shown in Fig. 7-8(b), $R_i = (3 + 20)\ k\Omega = 23\ k\Omega$.

We shall repeatedly use the resistance multiplication or transformation ability of the BJT. In Chapter 8 we shall use it to determine input resistance of a common collector circuit quickly and simply. It will be employed again in Chapter 10 to simplify analysis of multistage circuits.

7-9 Summary and Review

7-9.1 Summary.

1. After the dc bias currents and voltages are established, we use an ac model of both circuit and BJT to make a separate evaluation of ac signal currents and voltages superimposed on the dc biases.
2. The base and emitter terminals of a BJT act as if they present (a) a dc battery of $V_{BE} = 0.6\ V$ to the flow of dc bias current I_B, and (2) a resistance r_π to the flow of ac signal I_b.
3. To evaluate the ac model we need to know only (1) the dc bias current I_C from a dc analysis of the circuit, and (2) a value for β that is measured, obtained as the typical value from a data sheet, or estimated.
4. In real circuits an emitter resistor is usually added to stabilize the dc operating point. The emitter resistor also makes the gain constant, so gain depends not on the BJT but on R_L/R_E.
5. Input resistance, R_i depends on r_π, and if an emitter resistance is present, R_i is increased by βR_E.
6. When looking into the base terminal, any resistance in the emitter leg is transformed to appear as it it were multiplied by β.
7. There is a pattern that allows us to construct the voltage-gain equation. V_o/E_i equals

$$\text{current gain} \times \frac{\text{ac load resistance seen by } V_o}{\text{BJT input resistance seen by } E_i}$$

7-9.2 Review. The principles developed in this chapter will be reviewed and extended by the following examples.

Example 7-15: In Fig. 7-9 switch 1 is in the up position. $\beta = 100$. Find R_i and V_o/E_i for switch 2 in the (a) up position and (b) down position.
Solution: Evaluate r_π from Eq. (7-5a):

$$r_\pi = \frac{30\beta}{I_C} = \frac{30 \times 100}{1} = 3k\Omega$$

(a) From Fig. 7-8(b) and Eq. (7-12),

$$R_i = 3\ k\Omega + 100(1\ k\Omega) = 103\ k\Omega$$

From Eq. (7-14a),

$$\frac{V_o}{E_i} = \frac{5000}{1000} = 5$$

(b) With switch 2 down, C_E places an ac short across R_E. Then $R_i = r_\pi = 3\ k\Omega$ and R_E does not appear in the model, as in Fig. 7-6(b). From Eq. (7-9),

$$\frac{V_o}{E_i} = \beta \frac{R_L}{R_i} = 100 \frac{5000}{3000} = 166$$

Example 7-16: Repeat Example 7-15 except with switch 1 in the down position.
Solution: With switch 2 down, the ac load developing V_o is R_{ac} and consists of $R_L \| R'_L$. Evaluating,

$$R_{ac} = 5\ k\Omega \| 5\ k\Omega = 2.5\ k\Omega$$

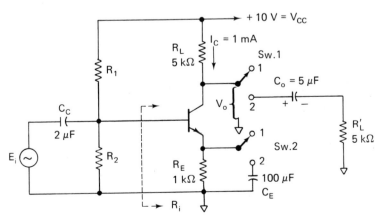

Figure 7-9 Circuit for Examples 7-15 to 7-17.

Now substituting R_{ac} for R_L in the gain equations of Example 7-15,

(a) $\dfrac{V_o}{E_i} = \dfrac{2500}{1000} = 2.5$ (b) $\dfrac{V_o}{E_i} = 100\dfrac{2500}{3000} = 83$

Example 7-17: In Fig. 7-9, what is the (a) dc collector voltage (not V_{CE}), (b) dc emitter voltage, and (c) V_{CE}?
Solution:

(a) The voltage drop across R_L is $I_C R_L = 1\ \text{mA} \times 5\ \text{k}\Omega = 5\ \text{V}$. Subtracting this drop from V_{CC} gives a collector voltage of $10 - 5 = 5\ \text{V}$.
(b) The voltage drop across R_E is $= I_C R_E = 1\ \text{mA} \times 1\ \text{k}\Omega = 1\ \text{V}$.
(c) Since both R_L and R_E drop 6 V from the 10-V supply V_{CC}, the remainder, 4 V, is V_{CE}.

Recall from Section 6-6.2 that C_o and C_E do not affect the dc operating point (once they are charged to V_C and V_E, respectively). Therefore, the results of Example 7-17 apply regardless of the position of either switch.

Problems

7-1 A transistor circuit has an input resistance of 2 kΩ. What current will it draw from a signal source of $E_i = 20\ \text{mV rms}$?

7-2 Evaluate g_m for a BJT operating at $I_C = 0.3\ \text{mA}$.

7-3 If dc collector current is doubled, is g_m doubled or halved?

7-4 If I_C is halved to 1 mA in Example 7-2, is r_π halved or doubled?

7-5 A BJT is biased with a base current of 10 μA or 0.010 mA. Find r_π.

7-6 If an operating point must be changed to increase r_π, should I_B be increased or decreased?

7-7 In Fig. 7-4, $R_L = 2.5\ \text{k}\Omega$, $R_B = 1.2\ \text{M}\Omega$, $V_{CC} = 12\ \text{V}$, $\beta = 50$. Find r_π.

7-8 Evaluate R_{in} for Problem 7-7.

7-9 Find the voltage gain for the circuit of Problem 7-7.

7-10 Does voltage gain increase or decrease when collector current I_C is increased?

7-11 If load resistor R_L is doubled, does voltage gain double or halve?

7-12 In Fig. 7-4, $R_L = 6.8\ \text{k}\Omega$ and $g_m = \frac{1}{100}$. Find voltage gain A_v.

7-13 I_C is changed to 1.5 mA in Fig. 7-6. Find A_V.

7-14 In Fig. 7-6, R_L is changed to 5 kΩ. Find A_V.

7-15 What changes result in Example 7-10 when we substitute a BJT with $\beta = 200$? Assume that I_C remains at 1 mA.

7-16 Evaluate voltage gain for Problem 7-15 by both approximate and exact methods.

7-17 If β is doubled, does A_V double in an amplifier with an unbypassed emitter resistor?

7-18 R_E is changed to 250 Ω in Example 7-10. Assume that I_C is unchanged. What are the new values of r_π and A_V?

7-19 In Fig. 7-7, $R_E = 250$ Ω, $R_L = 5$ kΩ, $V_{CC} = 20$ V, and $I_C = 1$ mA. Find peak output voltage V_{op}.

7-20 What changes result in Example 7-14 if a BJT with $\beta = 50$ is substituted?

7-21 What effect does a change in β to 50 have on Example 7-15? (I_C remains at 1 mA.)

8

Common-Collector and
Common-Base Circuits

8-0 Introduction

This chapter treats two other circuit configurations, common collector and common base. For each configuration we will study (1) biasing, (2) ac input resistance, and (3) voltage gain. Although each of these configurations finds limited use, together they form a versatile and high-performance circuit. When connected in this arrangement, the common-base circuit is the load for the common-collector circuit. This network is called a *differential amplifier*. Section 8-3 introduces the basic differential amplifier to illustrate an important and useful application for common-base and common-collector circuits.

8-1 Common-Collector Amplifier

In Fig. 8-1, ac input signal E_i is applied between base and collector, and output voltage V_o is taken between emitter and collector. Since the collector is common to both input and output, this circuit is a *common collector* (CC). Remember that when trying to determine the type of configuration, think only of the ac circuit (treat capacitors and dc supplies as short circuits). With this in mind, the collector in Fig. 8-1 is, for ac considerations, connected to the common point. Load resistor R_L, in a common-collector circuit, is in the emitter leg. But before any ac input signal can be applied to an amplifier, it first must be properly biased to set the dc operating point.

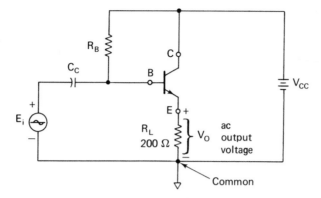

Figure 8-1 Common-collector circuit.

8-1.1 Biasing a Common-Collector Circuit. Biasing a common-collector circuit is identical to that of a common emitter. Therefore, equations developed in Chapter 6 to bias a common-emitter circuit will be used to bias a common collector. The following example illustrates the similarity.

Example 8-1: In the circuit of Fig. 8-2, determine the value of R_B to set the operating point for maximum possible output-voltage swing. $\beta = 50$, $V_{CC} = 20$ V.
Solution:

(1) From Section 6-5, set $V_{CE} = V_{CC}/2$ for maximum output-voltage swing. Thus

$$V_{CE} = \frac{20 \text{ V}}{2} = 10 \text{ V}$$

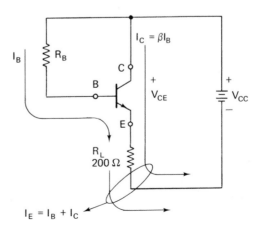

Figure 8-2 Biasing the common-collector circuit.

(2) The dc collector current may be found from

$$I_C \cong I_E = \frac{V_{CC} - V_{CE}}{R_L} \tag{8-1}$$

For Fig. 8-2,

$$I_C = \frac{20 \text{ V} - 10 \text{ V}}{200 \text{ } \Omega} = 50 \text{ mA}$$

(3) Writing the I_B loop equation to obtain base current I_B yields

$$V_{CC} = I_B R_B + V_{BE} + V_E \tag{8-2}$$

where V_E is the dc voltage across R_L:

$$\begin{aligned} V_E &= I_E R \cong I_C R_L \\ &= (50 \text{ mA})(200 \text{ } \Omega) = 10 \text{ V} \end{aligned} \tag{8-3}$$

Using Eq. (4-2), base current I_B is

$$I_B = \frac{I_C}{\beta} = \frac{50 \text{ mA}}{50} = 1\text{mA}$$

Now, substituting into Eq. (8-2),

$$20 \text{ V} = (1 \text{ mA})R_B + 0.6 \text{ V} + 10 \text{ V}$$

or

$$R_B = \frac{20 \text{ V} - 0.6 \text{ V} - 10 \text{ V}}{1 \text{ mA}} = \frac{9.4 \text{ V}}{1 \text{ mA}} = 9.4 \text{ k}\Omega$$

If only 10% resistors are used when building this circuit, a 10-kΩ resistor may be chosen for R_B.

8-1.2 Input Resistance of a Common-Collector Circuit. To construct an equivalent and simplified ac model of a common-collector circuit, we need to use the information developed in Sections 7-8 and 7-9. Looking into the base terminal, the input resistance R_i is r_π in series with β times R_L. In equation form,

$$R_i \cong r_\pi + \beta R_L \tag{8-4}$$

The input resistance of the circuit, R_{in}, includes the effect of the base resistance R_B:

$$R_{in} = R_B \| R_i \tag{8-5}$$

Figure 8-3 shows the equivalent ac model of Fig. 8-1.

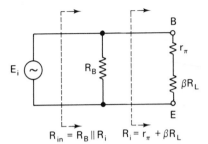

Figure 8-3 Ac input model of Fig. 8-1.

$R_{in} = R_B \| R_i \qquad R_i = r_\pi + \beta R_L$

Example 8-2: From Example 8-1 and Figs. 8-1 and 8-2, $\beta = 50$ and $R_B = 10 \text{ k}\Omega$. Find (a) r_π, (b) R_i, and (c) R_{in}.
Solution:

(a) From Eq. (7-5a) and Example 8-1,

$$r_\pi = \frac{30 \times \beta}{I_C \text{ (mA)}} = \frac{30 \times 50}{50} = 30 \ \Omega$$

(b) Using Eq. (8-4),

$$R_i = r_\pi + \beta R_L$$
$$= 30 \ \Omega + (50)(200 \ \Omega) = 30 \ \Omega + 10 \text{ k}\Omega = 10 \text{ k}\Omega$$

(c) Applying Eq. (8-5),

$$R_{in} = \frac{R_B R_i}{R_B + R_i} = \frac{(10 \text{ k}\Omega)(10 \text{ k}\Omega)}{10 \text{ k}\Omega + 10 \text{ k}\Omega} = 5 \text{ k}\Omega$$

The most significant lessons to be learned from Example 8-2 are: (1) input resistance, R_i, for the common-collector circuit is usually large because of the multiplication of βR_L; and (2) r_π in a common-collector circuit is almost always very small with respect to R_L.

Thus r_π can be neglected in Eq. (8-4) and the equation simplified to

$$R_i \cong \beta R_L \qquad \text{for } r_\pi \ll \beta R_L \tag{8-6}$$

8-1.3 Voltage Gain of a Common-Collector Circuit. When the ac input voltage, E_i, in Fig. 8-1 is applied to the base, an ac output voltage, V_o, is developed across load resistor R_L. To be able to apply the standard form for voltage gain, Eq. (7-10), we need an expression for current gain. For a common-collector circuit, the ac input current is I_b and the load or output current is the emitter current I_e. So current gain A_i for a common-collector circuit is

$$A_i = \frac{\text{ac emitter current}}{\text{ac base current}} = \frac{I_e}{I_b} = \frac{(\beta + 1)I_b}{I_b} = \beta + 1 \tag{8-7a}$$

However, $(\beta + 1)$ is very nearly equal to β and we can use the excellent approximation for A_i of a common-collector circuit as

$$A_i \cong \beta \qquad (8\text{-}7b)$$

Now apply the standard form for voltage gain, Eq. (7-10):

$$A_v = \frac{V_o}{E_i} = \text{current gain} \times \frac{\text{load resistance, } R_L}{\text{ac input resistance, } R_i}$$

Substituting values for the common-collector circuit yields

$$A_v = \frac{V_o}{E_i} = \beta \frac{R_L}{R_i} \qquad (8\text{-}8a)$$

But, from Eq. (8-6), $R_i \cong \beta R_L$ and, substituting into Eq. (8-8a), gives the simple but excellent approximation

$$A_v = \frac{V_o}{E_i} = \beta \frac{R_L}{\beta R_L} \cong 1 \qquad (8\text{-}8b)$$

To interpret Eq. (8-8b), visualize that ac input signal, E_i, between base and common terminal equals ac output voltage V_o between emitter and common terminal. Furthermore, when E_i goes positive, V_o goes positive, so there is no phase shift between input and output. Thus the ac emitter voltage always equals or *follows* the ac input voltage. This principle is recognized by the term *emitter–follower*, which is often applied to the common-collector configuration.

8-2 Common-Base Amplifier

In a *common-base* (CB) circuit (Fig. 8-4), ac input signal E_i is applied between emitter and base and ac output signal V_o is taken between collector and base. Capacitor C_B guarantees that the base, for ac purposes, is connected to the common point. As with common emitter and common collector circuits, our analysis of the common-base circuit begins with biasing.

8-2.1 Biasing a Common-Base Circuit. Figure 8-4 is the biasing network to establish the dc operating point and is the same circuit as Fig. 6-13. Therefore, we conclude that the biasing procedure is the same for both common-base circuits and common-emitter circuits with an emitter resistor. The bias procedure presented in Section 6-8.3 applies directly to Fig. 8-4.

Example 8-3: In Fig. 8-4, $V_{CC} = 20$ V, $R_L = 5$ kΩ, $R_E = 200$ Ω, and $\beta = 50$.

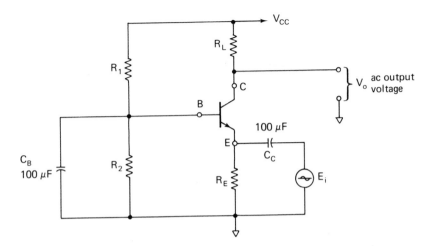

Figure 8-4 Common-base circuit.

Find R_1 and R_2 to set $I_C = 2$ mA; choose $R_B = R_2 = 10 \times R_E = 2$ kΩ.
Solution: The dc emitter voltage is

$$V_E = I_E R_E \cong I_C R_E = 2 \text{ mA} \times 200 \text{ Ω} = 0.4 \text{ V}$$

The dc base voltage V_B is

$$V_B = V_{BE} + V_E = 0.6 + 0.4 = 1.0 \text{ V}$$

From Eq. (6-15b),

$$V_B \cong V_{BB} = V_{CC}\frac{R_2}{R_1 + R_2}, \qquad 1.0 = 20\frac{2 \text{ kΩ}}{R_1 + 2 \text{ kΩ}} \quad \text{or} \quad R_1 = 38 \text{ kΩ}$$

8-2.2 Input Resistance of a Common-Base Circuit. In Fig. 8-4 capacitor C_C couples the ac signal from E_i directly to the emitter terminal and blocks the flow of dc current to E_i. Emitter resistor R_E is needed, or else E_i would be short-circuited to ground. For ac signals, capacitor C_B is large enough to place the base terminal at ac ground potential. Figure 8-5(a) is the ac model. Note that C_B and V_{CC} short-circuit R_1 and R_2 to ground. R_E is omitted for simplicity in the small-signal model of Fig. 8-5(a).

Emitter terminal E is the input terminal. The ac resistance seen looking into the transistor between emitter and ground is R_i. Define the input or ac resistance for the common-base circuit as

$$R_i = \frac{E_i}{I_i} \tag{8-9a}$$

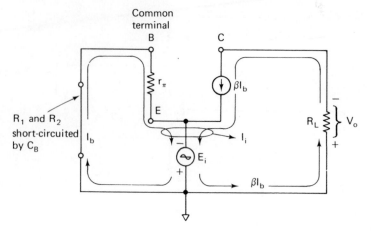

(a) Small-signal model of the CB circuit in
Fig. 8-4. R_E omitted for simplicity

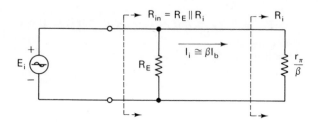

(b) Equivalent simplified ac input model of (a)

Figure 8-5 The common-base circuit of Fig. 8-4 is modeled in (a) and
simplified in (b).

where $I_i = I_b + \beta I_b = (1 + \beta)I_b \cong \beta I_b$. Then

$$R_i = \frac{E_i}{\beta I_b} \tag{8-9b}$$

From the I_b loop in Fig. 8-5(a) we see that $E_i = I_b r_\pi$ and, substituting into
Eq. (8-9b),

$$R_i = \frac{I_b r_\pi}{\beta I_b} = \frac{r_\pi}{\beta} \tag{8-9c}$$

Equation (8-9c) states that *the resistance in the base leg is divided by β when
looking into the emitter terminal.*

Figure 8-5(b) is a simplified model showing ac input resistance R_i and the total ac resistance, R_{in}, as seen by E_i.

$$R_{in} = R_E \| R_i \tag{8-10}$$

Example 8-4: In Fig. 8-4, $V_{CC} = 20$ V, $R_L = 5$ kΩ, $R_E = 200$ Ω, and $\beta = 50$. R_1 and R_2 are chosen to establish $I_C = 2$ mA. Find (a) r_π and (b) R_i and R_{in}.
Solution:

(a) From Eq. (7-5a),

$$r_\pi = \frac{30 \times 50}{2} = 750 \ \Omega$$

(b) From Eq. (8-9c),

$$R_i = \frac{r_\pi}{\beta} = \frac{750 \ \Omega}{50} = 15 \ \Omega$$

(c) From Fig. 8-5b,

$$R_{in} = R_E \| R_i = \frac{R_E R_i}{R_E + R_i} = \frac{200 \times 15}{200 + 15} = 14 \ \Omega$$

Example 8-4 shows that input resistance of the BJT for a common-base circuit is *extremely* low, and for all practical purposes, $R_{in} \cong R_i$.

8-2.3 Voltage Gain of a Common-Base Circuit. The standard format for voltage gain, Eq. (7-10), is applicable to common-base circuits.

$$A_v = \text{current gain} \times \frac{\text{load resistor, } R_L}{\text{input resistor, } R_i}$$

Current through R_L is βI_o and input current from E_i is $(\beta + 1)I_b$, so current gain A_i is

$$A_i = \frac{\text{output current}}{\text{input current}} = \frac{\beta I_b}{(\beta + 1)I_b} = \frac{\beta}{\beta + 1} \tag{8-11}$$

Recall from Eq. (4-6) that $\beta/(\beta + 1) = \alpha \cong 1$. Now, applying the standard format with R_i given by Eq. (8-9c),

$$A_v = \frac{\beta}{\beta + 1} \frac{R_L}{r_\pi/\beta} \cong \frac{\beta R_L}{r_\pi} \tag{8-12}$$

Canceling the $(\beta + 1)$ and β terms in the denominator of Eq. (8-12) gives Eq. (7-11), which is the same as that for a common-emitter circuit without an emitter resistor.

Example 8-5: Calculate voltage gain for the circuit of Example 8-4.
Solution: Applying Eq. (8-12),

$$A_v = (50)\frac{5 \text{ k}\Omega}{750 \ \Omega} = 333$$

8-3 Basic Differential Amplifier

One of the most practical circuits using both common-collector and common-base configurations is shown in Fig. 8-6. This circuit is used as the basic building block by integrated circuit manufacturers for use in audio, video, intermediate-frequency (IF), tuned, and operational amplifiers. Figure 8-6 shows that this circuit may have two input sources, E_1 and E_2. Output voltage may be taken from the collector of either transistor to ground, V_{o1}, or V_{o2}. We can also measure V_{oD}, which is the difference between V_{o1} and V_{o2}. This difference voltage gives the circuit the name *difference amplifier* or, more commonly, *differential amplifier*. As with any amplifying circuit, the first step is to bias it.

8-3.1 Biasing a Differential Amplifier. Biasing the differential amplifier is actually simpler than biasing one-transistor amplifiers. However, certain

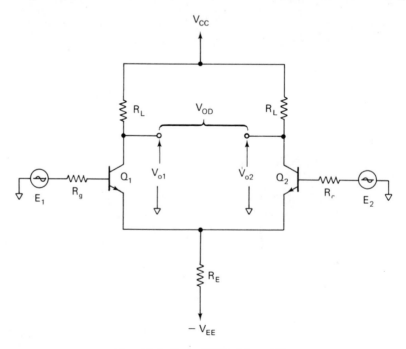

Figure 8-6 Basic differential amplifier.

steps or procedures should be followed: (1) Make the two dc supplies, V_{CC} and V_{EE}, equal; (2) choose both BJTs Q_1 and Q_2 with equal values of β and V_{BE}; (3) load resistors R_L should be equal, so use 1% resistors; (4) also use 1% resistors for R_g. In conclusion, the left side Q_1, R_L, and R_g is *matched* to the right side Q_2, R_L, and R_g, and the circuit is said to be *symmetrical*.

In most differential amplifiers, R_g usually does not exceed 1 kΩ and the base current through it is less than 10 μA. Therefore, the voltage drop across it is less than 0.01 V (1 kΩ × 10 μA = 0.01 V) and can be neglected.

Because of the symmetry of a differential amplifier, both dc collector currents are equal. As shown in Fig. 8-7, these collector currents flow through R_E. Thus

$$I_E = 2I_C \tag{8-13a}$$

or

$$I_C = \frac{I_E}{2} \tag{8-13b}$$

Now if we apply Kirchhoff's voltage law around the dotted path in Fig. 8-7,

$$V_{EE} = I_E R_E + 0.6 \text{ V}$$

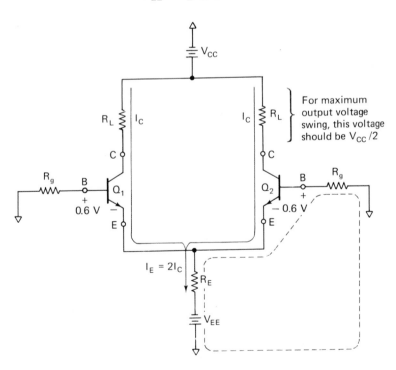

Figure 8-7 Bias currents for the basic differential amplifier.

Solving for I_E,

$$I_E = \frac{V_{EE} - 0.6\text{ V}}{R_E}$$

Normally V_{EE} is much greater than 0.6 V, so

$$I_E \cong \frac{V_{EE}}{R_E} \tag{8-14}$$

Equation (8-14) states that if V_{EE} is specified, R_E may be chosen to set I_E. From Eq. (8-13b), I_C equals $I_E/2$; therefore, the BJT's collector current is established by V_{EE} and R_E. Rewriting Eq. (8-13b) so that I_C is calculated directly,

$$I_C = \frac{V_{EE}}{2R_E} \tag{8-15}$$

When biasing any transistor amplifier we usually choose an operating point that gives maximum output voltage swing. For the differential amplifier, maximum output voltage is obtained when $V_{CE} = V_{CC}/2$, or $I_C = V_{CC}/2R_L$ and $R_L = R_E$. The simplicity of biasing the differential amplifier is illustrated in the next example.

Example 8-6: In Fig. 8-7, $V_{CC} = V_{EE} = 15$ V, $R_L = R_E = 7.5$ kΩ, and β for both transistors is 100. Estimate the operating point (a) collector current I_C, (b) base current I_B, and (c) collector-to-ground voltage V_{CG}.
Solution:

(a) Using Eq. (8-15),

$$I_C = \frac{V_{EE}}{2R_E} = \frac{15\text{ V}}{2(7.5\text{ k}\Omega)} = 1\text{ mA}$$

(b) $$I_B = \frac{I_C}{\beta} = \frac{1\text{ mA}}{100} = 10\ \mu\text{A}$$

(c) $$V_{CC} = I_C R_L + V_{CG}$$

or $$V_{CG} = V_{CC} - I_C R_L = 15\text{ V} - (1\text{ mA})(7.5\text{ k}\Omega) = 7.5\text{ V}$$

(*Note:* Since $V_{CG} = 7.5$ V and $V_{CC} = 15$ V, the dc voltage across R_L must also be 7.5 V. We conclude that for maximum output-voltage swing, V_{CC} divides equally between R_L and V_{CG}.)

In summary, make R_L equal to R_E and V_{EE} equal to V_{CC}. This automatically establishes an operating point for maximum output-voltage swing at $I_C = V_{EE}/2R_E$.

8-3.2 Input Resistance of a Differential Amplifier. The ac input resistance seen by E_i in Fig. 8-6 is expressed simply as

$$R_{in1} = 2(r_\pi + R_g) \tag{8-16}$$

To learn how Eq. (8-16) is obtained, we begin at the emitter of Q_2 and work our way back to the base of Q_1. A step-by-step breakdown is illustrated in Fig. 8-8. Q_2 is a common-base circuit and the resistance seen looking into

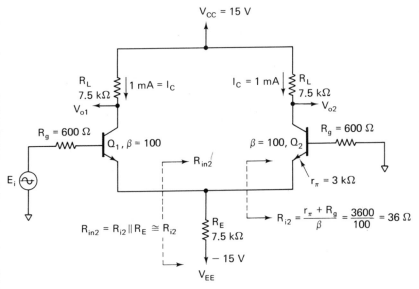

(a) Differential amplifier circuit to find
input resistance presented to E

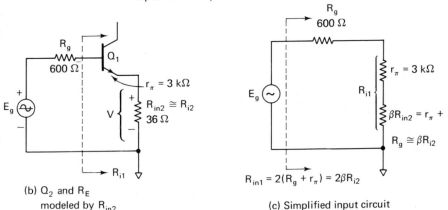

(b) Q_2 and R_E
modeled by R_{in2}

(c) Simplified input circuit

Figure 8-8 The equivalent input resistance of Q_2 in (a) is modeled by R_{in2} in (b) and simplified further in (c).

the emitter of Q_2 is

$$R_{i2} \cong \frac{r_\pi + R_g}{\beta} \tag{8-17}$$

Evaluating r_π from Eq. (7-5a),

$$r_\pi = \frac{30 \times 100}{1} = 3 \text{ k}\Omega$$

and
$$R_{i2} = \frac{3000 + 600}{100} = 36 \ \Omega$$

which shows that R_{i2} is very small. As shown in Fig. 8-8(a), R_{in_2} is the parallel combination of R_E and R_{i2}:

$$R_{in2} = R_E \| R_{i2} = \frac{R_E R_{i2}}{R_B + R_{i2}} \tag{8-18}$$

Substituting values,

$$R_{in2} = \frac{(7.5 \text{ k}\Omega)(36 \ \Omega)}{7.5 \text{ k}\Omega + 36 \ \Omega} \cong 36 \ \Omega$$

Therefore, R_{in2} is essentially equal to R_{i2}. Figure 8-8(b) illustrates that R_E and the input resistance of Q_2 are modeled by the 36-Ω resistor in the emitter leg of Q_1. The input resistance seen by E_1 in Fig. 8-8 is

$$R_{in_1} \cong R_g + r_\pi + \beta R_{in2}$$
$$= 600 + 3000 + (100)(36) = 7200 \ \Omega \tag{8-19}$$

This is the same answer as that which would be obtained by Eq. (8-16):

$$R_{in1} = 2(600 + 3000) = 7200 \ \Omega$$

The essentials of understanding this input resistance can be summarized directly and simply. R_g and r_π of both transistors are equal, owing to symmetry. R_g and r_π of Q_2 are first divided by β of Q_2 and then multiplied by β of Q_1. Since the βs are equal, r_π and R_g of Q_2 appear to E_1 as unchanged and in series with r_π and R_g of Q_1.

8-3.3 Voltage Gain of a Differential Amplifier. Ac output voltage V_{o1} in Fig. 8-6 is measured from the collector of Q_1 to ground. Since E_1 is applied to the base of Q_1 and V_{o1} is measured from the collector, Q_1 acts as a common-emitter circuit. Writing the gain equation using the standard form, Eq. (7-10), and with R_i given by Eq. (8-16), we obtain

$$\frac{V_{o1}}{E_1} = \beta \frac{R_L}{2(r_\pi + R_g)} \tag{8-20}$$

V_{o1} goes negative when E_1 goes positive, so there is a 180° phase shift, as with any common-emitter circuit.

Output voltage V_{o2} is measured from the collector of Q_2 to ground. To obtain V_{o2} we need to obtain the ac voltage from the emitter of Q_1 and then from the collector of Q_2. Input voltage E_1 is applied to the base of Q_1, and voltage V in Fig. 8-8(b) is taken from its emitter and applied to the emitter of Q_2. Since we are measuring V from the emitter of Q_1, Q_1 acts as a common collector whose gain is

$$\frac{V}{E_1} = \beta \frac{R_{i2}}{2(\beta R_{i2})} = \frac{1}{2} \tag{8-21}$$

Note that R_{i2} is the ac load on Q_1 when the voltage is measured from emitter to ground. $2R_{i2}$ is the input resistance given by Fig. 8-8(c).

Voltage $V = E_1/2$ is applied to the emitter of Q_2 and the output voltage V_{o2} is taken from collector of Q_2. Therefore, Q_2 is a common-base circuit. The gain of Q_2 is

$$\frac{V_{o2}}{V} = \beta \frac{R_L}{R_{i2}} = \frac{\beta R_L}{r_\pi + R_g} \tag{8-22}$$

The gain of the differential amplifier is found by multiplying Eq. (8-21) by Eq. (8-22) to get

$$\frac{V_{o2}}{E_1} = \frac{\beta R_L}{2(r_\pi + R_g)} \tag{8-23}$$

There is no phase shift in either a common-collector or a common-base circuit, so V_{o2} is in phase with E_1. Comparing Eq. (8-20) with Eq. (8-23), we see that V_{o2} is equal in magnitude but 180° out of phase with V_{o1}.

Example 8-7: Using the values in Fig. 8-8(a), calculate V_{o2}/E_1.
Solution: Using Eq. (8-23) and r_π from Section 8-4.2,

$$\frac{V_{o2}}{E_1} = \frac{100(7.5)}{2(3000 + 600)} = 104$$

Problems

8-1 In Fig. 8-1, $R_L = 100 \ \Omega$, $\beta = 100$, $V_{CC} = 20$ V. Choose R_B to set $V_{CE} = 10$ V.

8-2 In Fig. 8-1, $R_B = 22 \ k\Omega$, $R_L = 500 \ \Omega$, and $\beta = 100$. Find I_C.

8-3 $\beta = 100$, $R_B = 10 \ k\Omega$, $R_L = 100 \ \Omega$, and $I_C = 100$ mA in Fig. 8-1. For the model of Fig. 8-2, find (a) r_π; (b) R_i; (c) R_{in}.

8-4 If β is changed to 100 in Example 8-2 and I_C remains at 50 mA, what changes occur in the solutions?

8-5 What is the current gain from base to emitter of a common-collector circuit with a BJT of $\beta = 60$?

8-6 Approximately what voltage gain is obtained from a common-collector circuit?

8-7 Does an emitter follower give more voltage gain than a common-emitter circuit?

8-8 Recalculate R_1 for Example 8-4 to set $I_C = 1$ mA. Keep $R_2 = 2$ kΩ.

8-9 What changes result in Example 8-4 when a BJT with $\beta = 100$ is substituted?

8-10 Compare the current gains of a BJT in a common-base and a common-emitter circuit.

8-11 In Fig. 8-4, $R_L = 3.3$ kΩ, $\beta = 50$, and $r_\pi = 1$ kΩ. Find A_V.

8-12 R_L and R_E are changed to 10-kΩ resistors in Example 8-6. Find (a) I_C; (b) I_B; (c) V_{CG}.

8-13 What input resistance R_{in_1} results if BJTs with $\beta = 50$ are substituted in the circuit of Fig. 8-8(a)?

8-14 Find V_{o1}/E_1 and V_{o2}/E_1 for Problem 8-13.

9

Power in Amplifiers

9-0 Introduction

In this chapter we again want to analyze the output circuit, but this time with respect to power. The questions to be answered are: (1) What power is dissipated by the transistor and load under bias conditions (no input signal) and with an input signal applied? (2) What power is supplied by the voltage source, V_{CC}, with and without an input signal?

Until now all our circuits would be classified as class A in operation. *Class A* circuits are so grouped because the output wave form is similar to the input wave form multiplied by the circuit's voltage gain. However, class A circuits waste power when no ac signal is applied. This waste is not significant for small-signal amplifiers that produce 1 W or less. When the output stage of an amplifier must produce several watts to drive a speaker, the wasted power has only disadvantages: (1) it is costly to produce, and (2) the heat generated must be dissipated. To overcome these problems, we shall analyze a complementary symmetry output stage. This stage consists of two power transistors, one for positive input signals, the other for negative signals. The advantage is that only when the transistor is conducting an ac input signal must the transistor dissipate power. With no input signal, both transistors are off and will dissipate only a very small amount of power. This type of operation is called *class B*. To clarify power relationships, we begin with a study of power dissipations in a resistor.

143

9-1 Basic Power Relationships

9-1.1 Resistor Power. From power fundamentals, the power developed in a load resistor, R_L, due to dc current I_C, Fig. 9-1(a), is equal to R_L times the square of the current. Using P_{RD} to signify resistor power due to the dc current, we have

$$P_{RD} = I_C^2 R_L \qquad (9\text{-}1a)$$

In Fig. 9-1(b) an ac current with peak value I_{cp} has an rms value of $I_c = 0.707 I_{cp}$. Calling power developed in load resistor R_L due to the ac current P_{RA},

$$P_{RA} = I_c^2 R_L \qquad (9\text{-}1b)$$

In the load resistor of a CE circuit, a dc bias current I_C flows in the absence of a signal. The resulting power is evaluated from Eq. (9-1a) and, in an amplifier circuit is not used and so is wasted. When a signal voltage is connected to the CE circuit, an ac component of collector current is superimposed on the dc bias current as shown in Fig. 9-1(c). The resulting resistor power P_R is *increased* to

$$P_R = P_{RD} + P_{RA} \qquad (9\text{-}1c)$$

In an amplifier, the only useful component is the ac collector current.

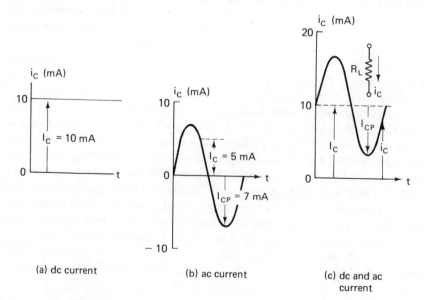

(a) dc current

(b) ac current

(c) dc and ac current

Figure 9-1 Ac and dc power components in a resistor.

The output power of the ac signal is designated by P_o and is equal to P_{RA}:

$$P_o = P_{RA} \qquad (9\text{-}1\text{d})$$

9-1.2 Supply Power. Power supply V_{CC} supplies power P_s to both transistor and load resistor R_L in Fig. 9-2. Since base current is much smaller than collector current ($I_B = I_C/\beta$), we neglect the tiny base-circuit power in all that follows. V_{CC} is a dc voltage, and for all practical purposes, its internal resistance is considered negligible and power supplied by it is

$$P_s = P_R + P_D \qquad (9\text{-}2)$$

In a BJT amplifier circuit the current drawn from V_{CC} will have a wave shape similar to Fig. 9-1(c). That is, current furnished by the power supply has an ac signal component riding on a dc bias component. The power supplied from a dc voltage source that furnishes a dc plus an ac current is not covered in the literature that deals with electrical fundamentals. The reason is that no such combination exists with passive elements such as resistors, capacitors, and inductors. To find the *average* power supplied by V_{CC} we reason that its average voltage is simply V_{CC}. In Fig. 9-1(c), the *average* value of the complex current wave i_C is seen to be equal to its dc component I_C. Note carefully that as long as the positive and negative peak values of I_{cp} are equal, the average current is I_C. Thus average supply power, P_s, is constant and does *not* depend upon how much ac signal current it supplies but only on the average dc current, or

$$P_s = V_{CC}I_C \qquad (9\text{-}3)$$

Example 9-1: In Fig. 9-2, $V_{CC} = 20$ V, $R_L = 1$ kΩ, and $I_C = 10$ mA. With no signal, $I_{cp} = 0$. Find the power (a) developed in load R_L and (b) supplied from V_{CC}; (c) account for any difference between (a) and (b).

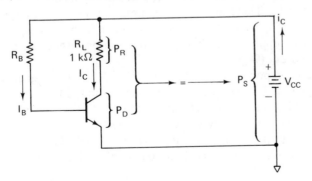

Figure 9-2 P_s is the average power supplied to load resistor R_L and the BJT.

Solution:

(a) From Eq. (9-1a), $P_{RD} = I_C^2 R_L = (0.010)^2 \times 1000 = 0.1$ W.
From Eq. (9-1b), $P_{RA} = I_c^2 R_L = (0)^2 R_L = 0$ W $= P_o$.
From Eq. (9-1c), $P_R = P_{RD} + P_{RA} = 0.1$ W $+ 0$ W $= 0.1$ W.
(b) From Eq. (9-3), $P_s = V_{CC}I_C = (20$ V$)(10$ mA$) = 0.2$ W.
(c) The difference between supplied power P_s and power P_R delivered to R_L must be in the transistor as P_D. From Eq. (9-2),

$$P_D = P_s - P_R = 0.2 \text{ W} - 0.1 \text{ W} = 0.1 \text{ W}$$

Example 9-2: Repeat Example 9-1, but with a signal of $I_{cp} = 10$ mA.
Solution:

(a) P_{RD} is unchanged at 0.1 W. Translate I_{cp} to I_c, $I_c = 0.707I_{cp} \cong 7$ mA for use in Eq. (9-1b). $P_{RA} = (I_c)^2 R_L = (0.007)^2 \times 1000 \cong 0.05$ W $= P_o$. From Eq. (9-1c), $P_R = 0.1$ W $+ 0.05$ W $= 0.15$ W.
(b) P_s is unchanged at 0.2 W.
(c) From Eq. (9-2), $P_D = P_s - P_R = 0.20$ W $- 0.15$ W $= 0.05$ W. The results of Examples 9-1 and 9-2 are tabulated in Table 9-1 to facilitate comparison.

Table 9-1 POWER COMPARISON WITH AND WITHOUT SIGNAL FOR EXAMPLES
9-1 AND 9-2

	P_s	=	P_R	+	P_D	P_o
Without signal	200 mW	=	100 mW	+	100 mW	0 W
With signal	200 mW	=	150 mW	+	50 mW	50 mW

9-1.3 Transistor Power. Transistor power, P_D, was introduced very briefly in Section 9-1.2 as the difference between supplied power P_s and load resistor power P_R. There are some revealing and perhaps unexpected conclusions concerning transistor power to be drawn from a study of Table 9-1. First note that supply power P_s is constant with or without a signal. This is expected from Eq. (9-3).

Load resistor power P_R increases by 50 mW when a signal is applied. Transistor power, P_D, *decreases* by the same amount, 50 mW, when a signal is applied. This same 50 mW is output power, P_o, that is equal to P_{RA}. It appears that 50 mW of power was transferred *from* the BJT to the load resistor as output power upon application of a signal. We conclude that the BJT operates cooler with a signal, at 50 mW, than it does without a signal, at 100 mW.

The subscript D was chosen for transistor power from the word "dissipate," to describe the fact that the BJT must dissipate heat. Power, P_D, heats the transistor. The BJT must dissipate its heat into its environment. This process will be covered in Chapter 16 under temperature limitations.

How transistor power, P_D, varies with operating-point location and output signal voltage will be the subject of Sections 9-2.1 and 9-2.2. For now we conclude that power is supplied from V_{CC} to load R_L and BJT. BJT power appears as heat and must be dissipated to the environment. Load power appears as heat for P_{RD}. P_{RA} or P_o also appears as resistor heat and is apparently transferred from the transistor. *Thus the worst transistor heating occurs at no-signal.*

9-2 Power, Output Voltage, and the Operating Point

9-2.1 Power and the Operating Point. Since the operating point determines both collector current I_C and collector–emitter voltage V_{CE}, it also determines the no-signal or dc BJT power, P_D. As concluded in Section 9-1.3, P_D is *maximum* at no-signal and is the product of operating point I_C and V_{CE}, or

$$\text{no-signal } P_D = I_C V_{CE} \tag{9-4}$$

To see how P_D depends on operating-point location we extend the dc load-line techniques introduced in Chapter 5. A dc load line is drawn in Fig. 9-3(b) for a 1-kΩ load line and a supply voltage of 20 V. Operating point O is located halfway up the load line. Recall that point O is actually located by its coordinates of collector bias current I_C and collector–emitter voltage V_{CE}. Their product is shown by Eq. (9-4) to equal P_D. But graphically, in Fig. 9-3(b), V_{CE} is the base and I_C the height of the cross-hatched rectangle. Since the area of a rectangle equals the base times the height, the cross-hatched rectangle gives a picture of P_D.

The value of P_D becomes smaller, as shown in Fig. 9-3(c) and (d), when operating point O is located above and below the center of the load line. It can be concluded the P_D is maximum when operating point O is located in the center of the load line.

Rectangles bordered with heavy lines in Fig. 9-3(b) to (d) portray supply power, P_s, as indicated by Eq. (9-3). P_s increases as operating point O moves up the load line. Unshaded rectangles show resistor power P_R because they are the difference between P_s and P_D. Like P_s, P_R also increases with higher values of operating-point collector current or higher locations of the operating point.

9-2.2 Power and Output-Voltage Swing. As shown in Section 6-6.5, maximum peak output voltage, max V_{op}, is determined by the location of the operating point. It is the smaller of V_{CE} or $I_C R_L$ and is maximum when

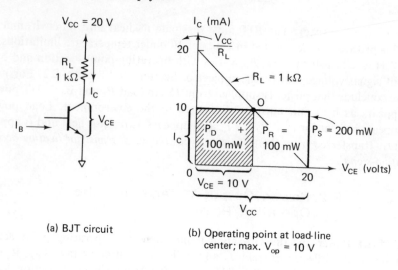

(a) BJT circuit

(b) Operating point at load-line center; max. $V_{op} = 10$ V

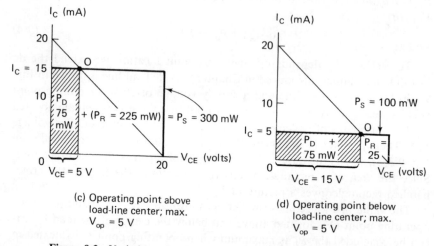

(c) Operating point above load-line center; max. $V_{op} = 5$ V

(d) Operating point below load-line center; max. $V_{op} = 5$ V

Figure 9-3 Variation of power levels and maximum possible output voltage with operating-point location.

the operating point is located at the center of the load line. Max V_{op} is shown to be 10 V for Fig. 9-3(b), 5 V for Fig. 9-3(c), and 5 V for Fig. 9-3(d). Thus the center-load-line location gives maximum possible output voltage and maximum transistor dissipation.

9-2.3 Summary and Conclusions. For maximum possible output power P_o, the operating point should be located at the center of the load line. Unfortunately, maximum transistor power dissipation, P_D, also occurs at this location. To conserve supply power, P_s, the operating point should be located as low as possible on the load line and still allow enough output voltage swing to develop the necessary output power. Since output power is determined by load resistance R_L or R_{ac}, and output voltage, we obtain the useful relationship of P_0 in terms of output voltage:

$$P_o = \left(\frac{V_{op}}{\sqrt{2}}\right)^2 \frac{1}{R_L} = \frac{V_{op}^2}{2R_L} \qquad (9\text{-}5)$$

9-3 Maximum-Efficiency Direct-Coupled Load

The amplifier of Fig. 9-3(a) is said to have a direct-coupled load because load, R_L, is directly connected to the BJT. Efficiency is the ratio of power output P_o to power input P_s and is represented by the Greek letter eta, η. Since P_s is constant for a given circuit, efficiency will depend on how much power output is obtained. P_o depends on how hard the amplifier is driven and will be a maximum when peak output voltage V_{op} is just equal to the maximum value allowed by the operating-point location. As noted in Section 9-2.2, the maximum possible V_{op} occurs with an operating point centered on the load line and equals one-half V_{CC}. Expressed mathematically,

$$\max V_{op} = \frac{V_{CC}}{2} \qquad (9\text{-}6a)$$

for $I_C = V_{CC}/2R_L$ and $V_{CE} = V_{CC}/2$.

Maximum output power is found by substituting for V_{op} from Eq. (9-6a) into Eq. (9-5):

$$\max P_o = \frac{V_{CC}^2}{8R_L} \qquad (9\text{-}6b)$$

From the definition of efficiency, η is found (in percent) by

$$\eta = \frac{P_o}{P_s} \times 100 \qquad (9\text{-}7a)$$

By combining Eqs. (9-6) and (9-3) with (9-7a), the maximum efficiency that can be obtained is given as

$$\eta_{\max} = 25\% \qquad (9\text{-}7b)$$

for the operating point located in the center of the load line and $V_{op} = V_{CC}/2$.

Example 9-3: In the circuit of Fig. 9-3(a), $V_{CC} = 12$ V and $R_L = 12\ \Omega$.

(a) Find operating-point coordinates for maximum possible ac output voltage. Find (b) P_s, (c) maximum V_{op}, (d) maximum P_o, and (e) maximum efficiency, in percent.

Solution:

(a) From Eq. (9-6a),

$$V_{CE} = \frac{V_{CC}}{2} = 6V, \qquad I_C = \frac{12 \text{ V}}{2 \times 12 \, \Omega} = 0.5 \text{ A}$$

(b) From Eq. (9-3), $P_s = V_{CC}I_C = 12 \text{ V} \times 0.5 \text{ A} = 6 \text{ W}$

(c) From Eq. (9-6a), $V_{op} = \dfrac{12 \text{ V}}{2} = 6 \text{ V}$

(d) From Eq. (9-6b), max $P_o = \dfrac{(12 \text{ V})^2}{8 \times 12 \, \Omega} = 1.5$ W. As a check from Eq.

(9-5), $P_o = \dfrac{(6 \text{ V})^2}{2 \times 12 \, \Omega} = 1.5 \text{ W}$

(e) From Eq. (9-7a),

$$\eta_{max} = \frac{P_o}{P_s} \times 100 = \frac{1.5 \text{ W}}{6.0 \text{ W}} \times 100 = 25\%$$

9-4 Maximum Output Power and Maximum BJT Dissipation for a Direct-Coupled Load

An extremely useful power relationship is the ratio of maximum output power to maximum BJT dissipation. If we know the power rating of a BJT from manufacturers' data sheets, we can determine the maximum output power that can be obtained from the BJT in a properly designed circuit. The relationship will be developed in an example.

Example 9-4: Find the ratio of maximum output power to maximum BJT dissipation.

Solution: For maximum output power the operating point must be centered at

$$I_C = \frac{V_{CC}}{2R_L} \quad \text{and} \quad V_{CE} = \frac{V_{CC}}{2}$$

Substituting into Eq. (9-4), we obtain maximum BJT dissipation, max P_D at the operating point:

$$\text{max } P_D = I_C V_{CE} = \frac{V_{CC}}{2R_L} \frac{V_{CC}}{2} = \frac{V_{CC}^2}{4R_L} \tag{9-8}$$

Taking the ratio of Eq. (9-6b) to Eq. (9-8),

$$\frac{\max P_o}{\max P_D} = \frac{1}{2} \tag{9-9}$$

The results of Example 9-4 tells us that for a direct-coupled load, the best we can get is 1W of output power for every 2 W of transistor power-handling capability.

Example 9-5: A transistor is rated at max $P_D = 1$ W. What is the maximum available output power?
Solution: From Eq. (9-9),

$$\max P_o = \tfrac{1}{2} \max P_D = \tfrac{1}{2} \times 1 \text{ W} = \tfrac{1}{2} \text{ W}$$

9-5 Power Relationship for an RC-Coupled Load

Load R_L is coupled by capacitor C to the BJT in Fig. 9-4(a). Output voltage V_o is developed across both R_C and R_L in Fig. 9-4(b). As shown in Fig. 9-4(c), the net ac load resistance that develops V_o is $R_{ac} = R_C \| R_L$. Maximum output voltage for an *RC*-coupled load is smaller than that for a direct-coupled load (see Section 6-6.3). Therefore, we would expect to obtain less power output and efficiency with *RC* coupling. The reductions in power and efficiency are investigated in an example.

Example 9-6: In Fig. 9-4(a), $R_C = R_L = 12\ \Omega$ and $V_{CC} = 12$ V. The operating point is located in the center of the *dc load line* at $I_C = 0.5$ A, $V_{CE} = 6$ V. Find (a) maximum V_{op}, (b) maximum output power delivered to load R_L (*not* to R_{ac}), (c) supply power, (d) maximum efficiency, and (e) P_{Dmax}.
Solution:

(a) From Eq. (6-7), maximum V_{op} is $V_{op} = I_c R_{ac} = 0.5$ A $\times$ $(12 \| 12)\ \Omega = 0.5$ A $\times$ $6\ \Omega = 3$ V.

(b) From Eq. (9-5), $P_o = \dfrac{V_{op}^2}{2R_L} = \dfrac{3 \times 3}{2 \times 12} = 0.375$ W

Note: Ac power delivered to R_{ac} is

$$\frac{V_{op}^2}{2R_{ac}} = \frac{(3 \text{ V})^2}{2 \times 6\ \Omega} = 0.75 \text{ W}$$

However, only the portion delivered to load R_L is *useful* and considered as output power.

(c) From Eq. (9-3), $P_s = V_{CC} I_c = 12$ V $\times$ 0.5 A $= 6$ W.

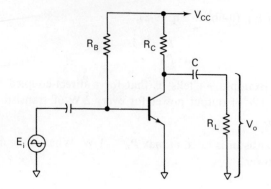

(a) Resistor-capacitor coupled load

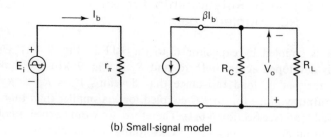

(b) Small-signal model

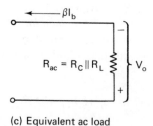

(c) Equivalent ac load

Figure 9-4 Circuit to determine power relations for *RC*-coupled load.

(d) Maximum efficiency, η, occurs at maximum V_{op}; from Eq. (9-7a),

$$\eta = \frac{P_o}{P_s} = \frac{0.375\ \text{W}}{6\ \text{W}} \times 100 = 6.25\%$$

(e) $P_{D\text{max}}$ occurs at no-signal and, from Eq. (9-4),

$$P_D = I_C V_{CE} = 0.5\ \text{A} \times 6\ \text{V} = 3\ \text{W}$$

We conclude from Example 9-6 that *RC*-coupled loads are very inefficient circuits for controlling power to a load. They are very costly because we must squander 6 W of supply power to get, at best, about $\frac{3}{8}$ W of useful output power. Also, the transistor must be rated at 3 W to obtain the $\frac{3}{8}$ W output. Thus for every watt of useful output, the transistor must be capable of dissipating 8 W, and V_{CC} must supply 16 W. *RC*-coupled loads have the advantage of simplicity, but in practice they are limited to delivering about 0.5 to 1 W of useful power.

The main disadvantage of these simple circuits is that the BJT conducts a large, wasteful bias current even when no signal is present. They waste power needlessly and thus pollute the environment. We turn our attention to a superior circuit arrangement that largely eliminates power waste when no signal is present.

9-6 Complementary Output Circuit

9-6.1 Description. A complementary output circuit is shown in Fig. 9-5. Complementary means that an *npn* BJT, Q_n, and a *pnp* BJT, Q_p, work together or complement one another. Three bias resistors, R_1, R_B, and R_2 are present to set the dc collector currents of Q_n and Q_p. Their operation will be covered in detail in Section 9-6.2. For now assume that the bias resistors ensure that Q_n and Q_p both conduct a tiny collector current of a few milliamperes when no signal is present ($E_i = 0$).

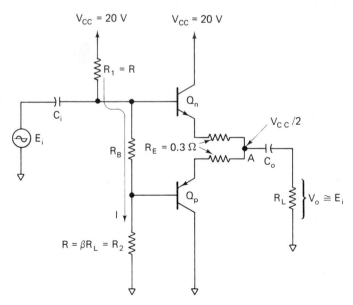

Figure 9-5 Complementary output circuit.

Input signal E_i is coupled by input capacitor C_i to the base of Q_n. R_B is a small-valued resistor of about 50 Ω, so E_i is also coupled to the base of Q_p. For positive-going half-cycles of E_i, the base–emitter junction of Q_n is forward-biased and Q_n turns on. Output signal is taken from the emitter of Q_n and coupled through C_o to load R_L. So Q_n acts as an emitter follower with a voltage gain of 1. During the positive half-cycle of E_i, the base–emitter junction of Q_p is reverse-biased and Q_p is turned off.

For negative half-cycles of E_i, Q_n is turned off and Q_p is turned on. Output voltage from the emitter of Q_p is applied to R_L, so Q_p also is an emitter follower. Thus Q_p and Q_n act like a single emitter follower, with Q_n conducting on positive-going E_i and Q_p conducting on negative-going E_i.

Output capacitor C_o is very large, of the order of 250 to 2000 μF. It should have a dc voltage rating larger than V_{CC}. C_o couples the signal voltage from the emitters of Q_n and Q_p to load R_L. Emitter resistors R_E are small in value and may be in the form of a fuse to protect the BJTs from conducting destructive currents. Resistor R_L represents a load, such as a speaker, that requires only ac currents.

9-6.2 Output Voltage and Power. Point A in Fig. 9-5 is biased to a dc voltage (with respect to ground) equal to one-half of V_{CC}. If Q_n is driven into saturation by a positive-going E_i, the collector–emitter voltage of Q_n will be only a few tenths of a volt (V_{CEsat}). So, neglecting the drop across R_E, point A will be driven positive to a voltage equal to V_{CC}. If E_i goes negative sufficiently to saturate Q_p, point A will be driven almost to ground potential. Thus voltage at point A can almost swing between limits of V_{CC} and ground for a maximum symmetrical peak output-voltage swing of $V_{op} = V_{CC}/2$.

Example 9-7: In Fig. 9-5, what are the approximate peak (a) output voltage swing and (b) current through R_L if $R_L = 10\ \Omega$? Neglect voltage drop across R_E. (c) What is the maximum power that can be delivered to R_L if E_i is a sine wave?
Solution:

(a) $V_{op} = V_{CC}/2 = 20\ \text{V}/2 = 10\ \text{V}$.
(b) Peak load current is $V_{op}/R_L = 10\ \text{V}/10\ \Omega = 1\ \text{A}$.
(c) Converting from peak to rms voltage, $V_o = 0.707 V_{op} = (0.707)\ (10\ \text{V}) = 7.07\ \text{V}$.

$$P_o = \frac{(V_o)^2}{R_L} = \frac{(7.07)^2}{10\ \Omega} = 5\ \text{W}$$

9-6.3 Biasing Theory. The complementary circuit of Fig. 9-5 is never found alone but is driven by other transistor circuits. An example will be

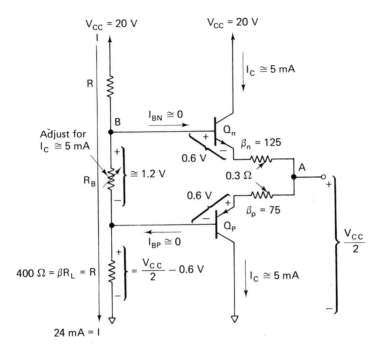

Figure 9-6 Dc bias currents and voltages for the complementary output stage.

presented in Chapter 10. Bias resistors R_1, R_2, and R_B are required if we wanted to test only the complementary output. But we can learn the basic bias requirements from the simplified dc model in Fig. 9-6.

The β of each transistor should be measured at a collector current equal to the peak expected load current. These values of β_n for Q_n and β_p for Q_p are averaged to obtain an average value for β. Equal resistors R are chosen from the nearest standard value equal to or less than βR_L. Selecting $R = \beta R_L$ ensures (1) that the BJT bases are biased at a dc voltage roughly equal to $V_{CC}/2$, and (2) that bias current, I, will allow enough base signal current to furnish full output voltage.

Resistor R_B is adjusted to put a slight forward bias on Q_n and Q_p so that they conduct a small bias collector current of approximately 5 mA. Q_n and Q_p are power BJTs that should be capable of conducting large collector currents exceeding 1 A (see Example 6-7). If R_B were not present, Q_n and Q_p would be cut off. All information in E_i between $+0.6$ V and -0.6 V would be lost at the output and introduce distortion. Since this distortion occurs where the signal crosses zero, it is called *crossover distortion*. Thus V_{CC}, R_1, and R_2 set the value of I, and R_B sets the 1.2-V forward bias on the BJTs. The technique is shown in an example.

Example 9-8: If $R_L = 4\,\Omega$, $\beta_n = 125$ and $\beta_p = 75$ in Fig. 9-5, find values of (a) R, (b) R_B, and (c) I.

Solution:

(a) Average β is found from

$$\beta = \frac{\beta_n + \beta_p}{2} = \frac{125 + 75}{2} = 100$$

Evaluate $R = \beta R_L = 100 \times 4\,\Omega = 400\,\Omega$. Select $R = 390\,\Omega$.

(b) From the voltage divider law, R_B is closely approximated from

$$R_B \approx 2R\frac{1.2}{V_{CC}} = 2(390)\frac{1.2}{20} = 47\,\Omega \tag{9-10}$$

Select R_B as a 50-Ω variable resistor, or substitute resistors for R_B of about 22, 33, and 47 Ω until I_C is a few milliamperes.

(c) From Ohm's law,

$$I = \frac{V_{CC}}{2R + R_B} = \frac{20\text{ V}}{(780 + 47)\Omega} = 24\text{ mA}$$

Note that base-bias currents are negligible with respect to I and do not affect I or the voltage levels. $I_{Bn} = I_C/\beta = 5\text{ mA}/125 = 40\ \mu\text{A}$, $I_{Bp} = 5\text{ mA}/75 = 67\ \mu\text{A}$.

9-6.4 Practical Biasing. When the complementary output stage is to be driven by another transistor circuit, R_1 must be removed, as in Fig. 9-7. Point B must be connected to a circuit that will maintain point B at approximately $V_{CC}/2$ while delivering 24 mA to terminal B. Under this condition bias currents and voltages will be the same in Figs. 9-6 and 9-7. Recall from Section 9-6.2 that point A had to be biased to $V_{CC}/2$ to realize maximum output-voltage swing. This bias voltage at point A is set by the voltage at point B because of the constant 0.6-V drop for V_{BE} of Q_n. (Voltage drop across the emitter resistors is negligible at $I_C R_E = 5\text{ mA} \times 0.3\,\Omega = 1.5\text{ mV}$.)

9-6.5 Summary. In the complementary output stage of Figs. 9-5 to 9-7, point A must be biased at one-half V_{CC} to realize a possible peak symmetrical load voltage swing of approximately $V_{CC}/2$ V. This is accomplished by a bias circuit that (a) sets point B at $(V_{CC}/2 + 0.6)$ V, (b) has a bias resistor R equal to βR_L, and (c) has a bias resistor R_B equal to $2.4R/V_{CC}$.

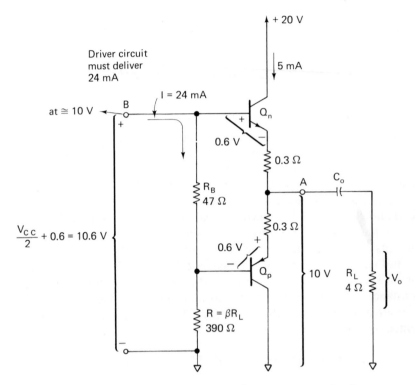

Figure 9-7 Practical biasing of complementary output circuit.

Maximum output power occurs when $V_{op} = V_{cc}/2$, and for Figs. 9-6 and 9-7,

$$\max P_o = \frac{V_o^2}{R_L} = \frac{(0.707 V_{op})^2}{R_L} = \frac{(0.707 \times 10)^2}{4} = 12.5 \text{ W} \qquad (9\text{-}11)$$

Bias power waste is drastically reduced to

$$P_s = V_{cc}I_c = 20 \text{ V} \times 5 \text{ mA} = 0.1 \text{ W}$$

The dramatic improvement in power waste is seen by comparing these values for P_o and P_s with Example 9-6.

Finally, maximum transistor dissipation will be one-fifth of max P_o, or

$$\max P_D = \frac{\max P_o}{5} \qquad (9\text{-}12)$$

and occurs when peak output voltage V_{op} is roughly equal to one-third of V_{cc}. Thus, for 12.5 W of maximum output power, BJTs must be able to dissipate 2.5 W each when the sinusoidal output voltage in Fig. 9-7 has a peak value of 8 V $= V_{op}$ and an rms value of $V_o = 5.6$ V.

9-7 Bootstrapping the Complementary Output Circuit

Bootstrapping is an exception to the rule that you cannot improve one aspect of circuit performance without incurring more expense or trading away some other aspect of circuit performance. By *bootstrapping*—changing the bottom connection of R in Fig. 9-7 from ground to the top of R_L—we double the input resistance. This process is shown in Fig. 9-8(a).

As far as dc operation is concerned, connecting $R_L = 4 \Omega$ in series with R has no effect on dc bias currents and voltages. If R_L is a speaker load, it will conduct 24 mA, but this is negligible with respect to expected signal currents of amperes and will not affect speaker operation.

In Fig. 9-8(b) a simplified ac signal model is shown for a nonbootstrapped circuit such as Fig. 9-7. Bias resistor R is in parallel with equivalent input resistance βR_L of either Q_n or Q_p (depending on which one is conducting for a positive or negative half-cycle of E_i). The resulting input resistance presented to E_i is

$$\text{nonbootstrapped } R_{\text{in}} \approx R \| \beta R_L = \beta R_L \| \beta R_L = \frac{\beta R_L}{2} \qquad (9\text{-}13)$$

A simplified ac model of the bootstrapped circuit in Fig. 9-8 is shown in Fig. 9-8(c) for positive-going E_i. Since Q_n acts as an emitter follower, output voltage V_o essentially equals input voltage E_i. (r_π is negligible for large signal currents.) Thus the signal voltage at one end of bias resistor R is about equal to the signal voltage at its other end. The signal-voltage drop across R is forced to be small, so very little signal current is conducted by R. In other words, the apparent resistance of R has become extremely large. We can think of R as an open circuit. It follows that resistance R_{in} presented to E_i is practically equal to R_i of the base terminal. As shown in Fig. 9-8(c),

$$\text{bootstrapped } R_{\text{in}} \approx R_i \approx \beta R_L \qquad (9\text{-}14)$$

By comparing Eqs. (9-13) and (9-14), we see that bootstrapping almost eliminates the effect of bias resistors on input resistance. Another advantage is that the dc input resistance of the complementary circuit R equals the ac bootstrapped input resistance βR_L. So the bootstrapped complementary circuit presents an equal dc and ac load to its driver circuit. This eliminates the reduction in output voltage swing that occurs when an ac load is smaller than the dc load.

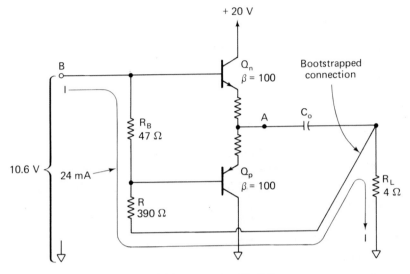

(a) Bootstrapping Fig. 9-7

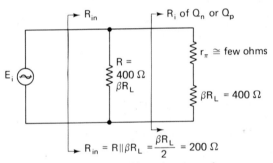

(b) Input resistance without bootstrapping

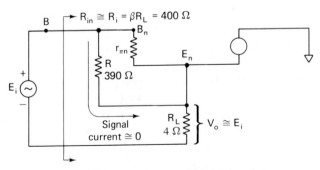

(c) Input resistance with bootstrapping

Figure 9-8 Boostrapping in (a) improves input resistance, as shown by (c).

Problems

9-1 Currents shown in Fig. 9-1 are conducted through a 100-Ω resistor. Find the resistor power developed by each current.

9-2 In Example 9-1, let the supply voltage and bias current be halved to $V_{CC} = 10$ V and $I_C = 5$ mA. Are the power levels also halved?

9-3 In Fig. 9-2, $V_{CC} = 24$ V, $I_C = 0.5$ A, and $R_L = 12\ \Omega$. No signal is present. Find P_R, P_s, and P_D.

9-4 A signal current is introduced in Problem 9-3 of $I_{cp} = 0.5$ A or $I_c = 0.35$ A. Find the new P_R, P_s, and P_D.

9-5 In a class A amplifier, does the transistor operate cooler with or without signal?

9-6 When a class A amplifier is biased for maximum output voltage swing, is it also biased for maximum power dissipation?

9-7 Which operating-point location uses less supply power, one located higher or lower on the load line?

9-8 What changes result in Example 9-3 when V_{CC} is increased to 24 V?

9-9 Find maximum transistor dissipation for Problem 9-8.

9-10 A BJT is rated at max $P_D = 10$ W. What is its maximum possible output power in a class A amplifier?

9-11 In Fig. 9-3(a), $V_{CC} = 24$ V and $R_L = 100\ \Omega$. Find (a) the operating point for maximum output; (b) P_s; (c) maximum V_{op}; (d) maximum P_o; (e) maximum P_D.

9-12 In Fig. 9-4(a), $R_C = R_L = 100\ \Omega$ and $V_{CC} = 24$ V. The operating point is located at $V_{CE} = 12$ V, $I_C = 0.12$ A. Find (a) maximum V_{op}; (b) maximum output power; (c) supply power; (d) maximum efficiency; (e) P_{Dmax}.

9-13 In Fig. 9-5, $V_{CC} = 24$ V and $R_L = 8\ \Omega$. Find maximum (a) peak output voltage; (b) peak load current; (c) power output.

9-14 In Fig. 9-6, $R_L = 8\ \Omega$ and $V_{CC} = 24$ V. Find (a) R; (b) R_B; (c) I.

9-15 If $V_{CC} = 24$ V in Fig. 9-7, what dc bias voltage should be established at point B with respect to ground?

9-16 BJTs with power ratings of $P_{Dmax} = 5$ W are used in the complementary output stage. What maximum output power can be obtained?

9-17 In Fig. 9-7, $R_L = 8\ \Omega$, $\beta = 100$, and $R = 800\ \Omega$. What is the input resistance?

9-18 When the circuit of Problem 9-17 is bootstrapped, what is the input resistance?

10

Multistage Amplifiers

10-0 Introduction

The term *stage* identifies a circuit arrangement of one BJT (or other active device) together with its associated bias network, power supply, and signal-coupling components. Multistage describes an interconnection of more than one stage whereby the output of one stage is connected to the input of another stage. This procedure is called *cascading* stages. Until now our primary concern has been to analyze the performance of a single stage. Unfortunately, performance of a single stage may be changed drastically when its output is wired to drive a second stage. We will learn in the next section how the performance of one stage is changed by cascading a second stage. We shall also learn how to analyze the overall performance of both stages.

10-1 Two-Stage BJT Amplifier

10-1.1 Multistaging for Voltage Gain. It is difficult to obtain a stable, large-voltage gain with a single amplifer stage. For example, we might try a CE circuit with an emitter resistor to establish a gain of 50 with $R_L/R_E = 50$. But R_E would necessarily be small. Then, if R_B was chosen as $10 \times R_E$ for stability, input resistance would be unacceptably small. Thus it is better to use two stages with smaller gains whose product will give the desired gain. Input resistance of the resulting first stage will be higher. However, as will be shown next, the second stage will unexpectedly load the first stage and change its gain.

161

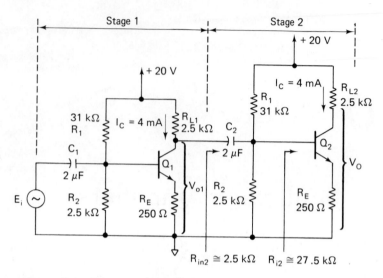

Figure 10-1 Two-stage amplifier made from cascading two identical stages.

10-1.2 Individual Stages. A two-stage amplifier is constructed by cascading two identical stages in Fig. 10-1. Each stage has identical load resistors R_{L1} and R_{L2} as well as identical emitter and bias resistors. If the stages were *not* cascaded by connecting the wire from capacitor C_2 to the collector of Q_1, each individual stage would exhibit the same performance as analyzed in the following example.

Example 10-1: Find the (a) input resistance R_{in}, (b) voltage gain, and (c) maximum possible output voltage for the Q_1 stage in Fig. 10-1. Assume that no connection is made between C_2 and collector of Q_1, and $\beta = 100$ for each BJT.

Solution:

(a) From Eq. (7-5a), $r_\pi = \dfrac{30\beta}{I_C} = \dfrac{30 \times 100}{4} = 750\ \Omega.$

From Eq. (7-12), $R_i \cong r_\pi + \beta R_E = 750\ \Omega + 100 \times 250\ \Omega = 25.7\ k\Omega.$

From Eq. (6-16a), $R_B \approx R_2 = 2.5\ k\Omega.$

From Eq. (7-13), $R_{in} = R_B \| R_i = 2.5\ k\Omega \| 25.7\ k\Omega \approx 2.5\ k\Omega.$

(b) From Eq. (7-14a), $A_V = \dfrac{V_{o1}}{E_1} \approx \dfrac{R_L}{R_E} = \dfrac{2.5\ k\Omega}{250\ \Omega} = 10$

(c) From Eq. (7-15a),

$$D = \frac{R_L}{R_L + R_E} = \frac{2500\ \Omega}{2500 + 250} = 0.9$$

$$V_{CE} = V_{CC} - I_C(R_L + R_E) = 20 - 0.004(2750) = 9\ \text{V}$$

$$\max V_{op} = DV_{CE} = 0.9 \times 9\ \text{V} \approx 8\ \text{V}$$

A similar analysis for the Q_2 stage would yield identical results.

From the results of Example 10-1 we conclude that each individual stage, when tested alone, has a voltage gain of 10, an input resistance of 2.5 kΩ, and can deliver a peak output voltage of about 8 V.

At this point it is tempting to draw the hasty but *erroneous* conclusion that the two cascaded stages would yield an overall gain of $10 \times 10 = 100$ from E_i to V_o. That is, 1 mV at E_i would be multipled by a Q_1 gain of 10 to give $V_{o1} = 10$ mV. Then V_{o1} would again be multiplied by a Q_2 gain of 10 to give $V_o = 100$ mV. However, this is *not* the case. As shown in the next section, a gain of Q_1 is reduced by the loading of Q_2, so the overall gain of the pair will be less than 100.

10-1.3 First-Stage Operation. The ac load resistance, R_{ac}, for Q_1 in Fig. 10-1 is not simply R_{L1}. It is the parallel combination of R_{L1} and R_{in2}, the input resistance to the Q_2 stage. Refer to the ac model of Fig. 10-2. Evaluating R_{ac} by modifying Eq. (6-5),

$$R_{ac} = R_{L1} \,\|\, R_{in2} = 2.5\ \text{k}\Omega \,\|\, 2.5\ \text{k}\Omega = 1.25\ \text{k}\Omega$$

we use Eq. (7-14a) to find the new gain for Q_1,

$$A_{V1} = \frac{V_{o1}}{E_1} \approx \frac{R_{ac}}{R_E} = \frac{1250\ \Omega}{250\ \Omega} = 5$$

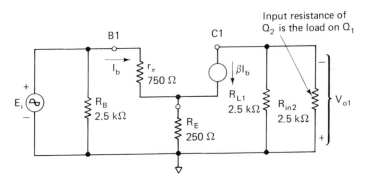

Figure 10-2 Ac model for Q_1 in Fig. 10-1.

We conclude that voltage gain of Q_1 is reduced from 10 to 5 when the connection is made from collector of Q_1 to C_2 in Fig. 10-1. Peak output-voltage swing is reduced from the 8-V value of Example 10-1(c) to a value found from Eq. (7-15b),

$$D_a = \frac{R_{ac}}{R_{ac} + R_E} = \frac{1250 \ \Omega}{(1250 + 250) \ \Omega} = 0.83$$

$$\max V_{op} = D_a I_C R_{ac} = (0.83)(0.004)(1250) \approx 4 \text{ V}$$

Thus connecting R_{in2} in parallel with R_{L1} halved the load resistance of Q_1 to halve *both* voltage gain and peak output-voltage swing.

10-1.4 Second-Stage Operation. Operation of the Q_2 stage is summarized by and identical to the performance found in Example 10-1. Its input voltage is V_{o1}, which is the output voltage of Q_1. V_{o1} will be amplified by Q_2's gain of 10, so V_o will equal $10V_{o1}$.

10-1.5 Two-Stage Operation. Operation of the two cascaded stages can be summarized from Figs. 10-1 and 10-2. E_i sees the 2.5 kΩ input resistance R_{in1} of Q_1 as found in Example 10-1. Voltage gain of Q_1 is $A_{V1} = 5$, from Section 10-1.3. Thus ac output voltage, V_{o1}, measured from collector of Q_1 to ground is

$$V_{o1} = A_{V1}E_i \qquad (10\text{-}1)$$

As shown in Fig. 10-1, V_{o1} is the ac input voltage to Q_2. V_{o1} is multiplied by the gain of Q_2, or A_{V2} to develop output voltage V_o. Thus

$$V_o = A_{V2}V_{o1} \qquad (10\text{-}2)$$

Substituting for V_{o1} from Eq. (10-1) into Eq. (10-2) gives overall output voltage V_o in terms of input voltage E_i, or

$$V_o = A_{V1}A_{V2}E_i \qquad (10\text{-}3a)$$

Overall two-stage gain, A_V, is therefore

$$A_V = \frac{V_o}{E_i} = A_{V1}A_{V2} \qquad (10\text{-}3b)$$

Example 10-2: Find the overall gain for the amplifier in Fig. 10-1.
Solution: From Section 10-1.3, $A_{V1} = 5$. From Section 10-1.2 and Example 10-1, $A_{V2} = 10$. From Eq. (10-3b),

$$A_V = \frac{V_o}{E_i} = 5 \times 10 = 50$$

Example 10-3: If $E_i = 10$ mV in Example 10-2, find (a) V_{o1} and (b) V_o.
Solution:

(a) From Eq. (10-1), $V_{o1} = A_{V1}E_i = 5 \times 10$ mV $= 50$ mV

(b) From Eq. (10-2), $V_o = A_{V2}V_{o1} = 10 \times 50$ mV $= 500$ mV, or from Eq. (10-3a), $V_o = A_{V1}A_{V2}E_i = 5 \times 10 \times 10$ mV $= 500$ mV

10-2 Loading Effects

A typical audio amplifier requires an input signal of about 0.2 V rms, amplifies it with a gain of 25 to 30 and applies the amplified voltage to a speaker. With the output switch of Fig. 10-3 in the up position, the Q_1 stage would provide a gain of $R_C/R_E \approx 30$ and an input resistance of $R_{in2} \approx R_2 = 2200$ Ω. However, when we threw the switch to its down position the ac load on Q_1 would be reduced to the 8-Ω speaker load R_L, because $R_{ac} = R_C \| R_L \approx R_L$. A voltage gain of Q_1 would now be $R_{ac}/R_E \approx 0.04$. V_o would equal 0.2 V $\times$ 0.04 $= 80$ mV and be insufficient to drive the speaker.

The failure of the circuit in Fig. 10-3 to perform properly is the result of the drastic reduction in R_{ac} of Q_1 because of the low resistance of R_L. What is needed is a circuit modification that will multiply the low resistance of R_L to a value large enough to have a minimum effect on R_{ac} of Q_1. For example, if R_L were made to appear as a 68-kΩ resistor or 10 times R_C, then R_{ac} of Q_1 would remain about equal to R_C. Then gain of Q_1 would stay at 30. Recall that the emitter-follower or common-collector circuit multiplied emitter-leg resistance by β. Therefore, we should look for a circuit that places R_L in the emitter leg of a common-collector configuration. But one transistor multiplication is not enough. Assuming a β of 100, R_L would be multiplied to $100 \times 8 = 800$ Ω. With 800 Ω paralleling R_C in Fig. 10-4, R_{ac} would still be

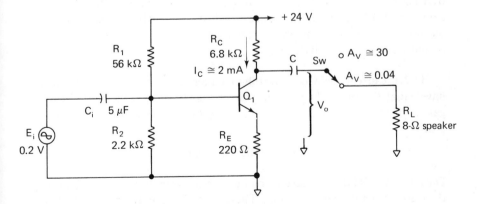

Figure 10-3 Circuit to demonstrate need for multistages.

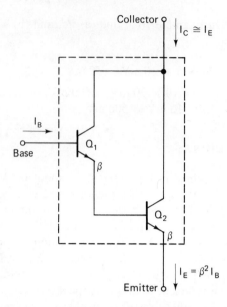

Figure 10-4 Darlington pair constructed from two *npn* transistors.

too small and reduce Q_1's gain to $800/220 \approx 4$. We need two common-collector circuits in cascade so that β of both transistors will multiply R_L. Thus R_L would appear as $100 \times 100 \times 8$, or $80,000 \ \Omega$ in parallel with R_C and have negligable effect on gain of Q_1. The cascade transistors are called a Darlington pair and are studied in Section 10-3.

10-3 Darlington Pair

The combination of a common collector driving a common collector as shown in Fig. 10-4 is called a *Darlington pair*. The advantage is that the transistors act as a single unit yielding (1) extremely large values of current gain, and (2) high input resistance. Each of these advantages, along with biasing a Darlington pair and obtaining the voltage gain of a Darlington amplifier, will now be discussed.

10-3.1 Current Gain of a Darlington Pair. Assume that the transistors have equal β_s, for the Darlington pair of Fig. 10-4; then the overall current gain is β^2 ($\beta \times \beta$). If each transistor has a β of 100, the current gain for the pair is

$$\beta^2 = (100)^2 = 10,000$$

Darlington pairs do not have to be built using discrete components because they are available commercially. These units have both transistors manufactured on a single chip, which virtually guarantees equal βs and current gains greater than 10,000. Note in Fig. 10-4 that only three leads are brought out, the equivalent of base, emitter, and collector.

For any transistor, emitter current is approximately current gain times base current. Therefore, for a Darlington pair

$$I_E \approx \beta^2 I_B \qquad (10\text{-}4)$$

Example 10-4: Calculate base current I_B in Fig. 10-4 if $I_E = 1$ A, and both transistors have βs equal to 100.
Solution: Rearranging Eq. (10-4),

$$I_B = \frac{I_E}{\beta^2} = \frac{1 \text{ A}}{10,000} = 0.1 \text{ mA}$$

Thus only a small value of base current is required to produce a large value of emitter current.

10-3.2 Biasing a Darlington Pair. Biasing a Darlington amplifier is similar to that of a single transistor common-emitter or common-collector stage except that now two base-to-emitter voltages must be taken into consideration. Writing a loop equation for the circuit of Fig. 10-5 to include V_{CC}, V_{BE}, plus the voltage drops across R_B and R_E, we obtain

$$V_{CC} = I_B R_B + 2V_{BE} + I_E R_E \qquad (10\text{-}5)$$

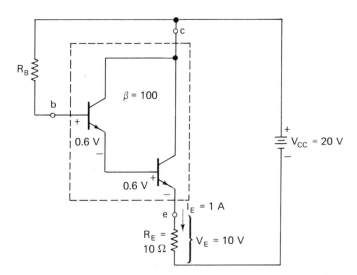

Figure 10-5 Bias arrangement for the Darlington circuit used in Example 10-5.

Solving for R_B yields

$$R_B = \frac{V_{CC} - 2V_{BE} - I_E R_E}{I_B} \tag{10-6}$$

Example 10-5: Determine the value of bias resistor R_B in Fig. 10-5 to produce an emitter current of 1 A. Both transistors are silicon, with $V_{BE} = 0.6$ V.
Solution: From Eq. (10-4) and Example 10-4,

$$I_B = \frac{I_E}{\beta^2} = \frac{1 \text{ A}}{10,000} = 0.1 \text{ mA}$$

and the voltage drop

$$I_E R_E = (1 \text{ A})(10 \text{ } \Omega) = 10 \text{ V}$$

Applying Eq. (10-6),

$$R_B = \frac{(20 - 2 \times 0.6 - 10) \text{ V}}{0.1 \text{ mA}} = \frac{8.8 \text{ V}}{0.1 \text{ mA}} = 88 \text{ k}\Omega$$

10-3.3 Input Resistance of a Darlington Pair. When an ac signal is applied to an amplifier, it is necessary to know the value of input or ac resistance of the transistor stage. This is because input resistance loads down (draws an excessive amount of current from) the preceding stages. Input resistance of a Darlington pair is very high, because the resistance "seen" looking into the base equals β^2 times any resistance connected in the emitter lead. In equation form,

$$R_i = \beta^2 R_E \tag{10-7}$$

Example 10-6: Calculate input resistance of the Darlington pair, R_i, for circuit values given in Fig. 10-6(a).
Solution: Applying Eq. (10-7),

$$R_i = (100)^2(10 \text{ } \Omega) = 100 \text{ k}\Omega$$

(*Note:* The low value of $R_E = 10$ Ω appears as 100 kΩ "seen" looking into base. This high input resistance is the major advantage of a Darlington pair.)
Input resistance R_i, however, is not the input resistance of the stage. Figure 10-6(b) is an equivalent input resistance circuit and shows that the stage resistance (or input resistance of a Darlington amplifier) is R_{in}, where

$$R_{in} = R_B \| R_i \tag{10-8}$$

Example 10-7: Determine the input resistance of the Darlington amplifier of Fig. 10-6(a).

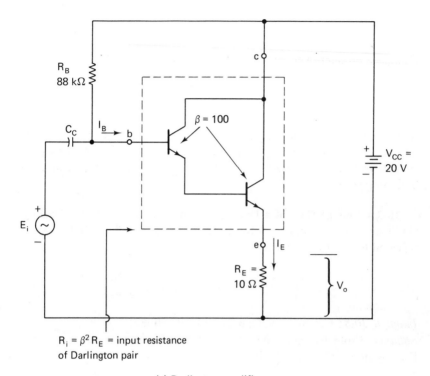

$R_i = \beta^2 R_E$ = input resistance
of Darlington pair

(a) Darlington amplifier

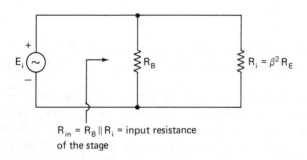

$R_{in} = R_B \| R_i$ = input resistance
of the stage

(b) Equivalent input resistance circuit

Figure 10-6 Determining the input resistance of a Darlington amplifier.

Solution: From Eq. (10-8),

$$R_{in} = 88 \text{ k}\Omega \| 100 \text{ k}\Omega$$

Using the product over the sum rule,

$$R_{in} = \frac{(88 \text{ k}\Omega)(100 \text{ k}\Omega)}{88 \text{ k}\Omega + 100 \text{ k}\Omega} = 46.8 \text{ k}\Omega$$

As with other amplifying stages, the overall input resistance is limited by the biasing resistor. Although R_{in} is approximately half of R_i, R_{in} is usually large enough so that a Darlington amplifier does not significantly load down a preceding stage.

10-3.4 Voltage Gain of a Darlington Pair. Voltage gain of a Darlington amplifier may be found from the general expression of Eq. (7-10), which is repeated here for convenience:

$$A_v = \frac{V_o}{E_i} = \text{current gain} \times \frac{\text{ac load resistance}}{\text{input resistance, } R_i}$$

Example 10-8: Calculate voltage gain of Fig. 10-6(a).
Solution: From Example 10-6, $R_i = 100 \text{ k}\Omega$. Recall that current gain of a Darlington amplifier is β^2; therefore,

$$A_v = (100)^2 \times \frac{10 \text{ }\Omega}{100 \text{ k}\Omega} \cong 1$$

Thus the voltage gain of the Darlington amplifier in Fig. 10-6(a) is no greater than the voltage gain of a single-stage common-collector circuit. However, the purpose of using a Darlington amplifier is not voltage gain but rather high input resistance.

10-4 Multistage Amplifier with Complementary Output

10-4.1 Introduction. Recall from Section 9-6 that the complementary stage acts as an emitter follower and has the distinct advantage of wasting very little power when no ac signal is present. However, the disadvantage is that a low value of load is usually connected to this type of stage such as a 4- or 8-Ω speaker. Input resistance (βR_L) is too low to be connected directly to a voltage-gain stage. To reduce this loading, the solution is a Darlington pair. The complementary output stage (Q_n and Q_p) of Fig. 10-7 is similar to Example 9-8. In Fig. 10-7, Q_D has been inserted between the gain stage and

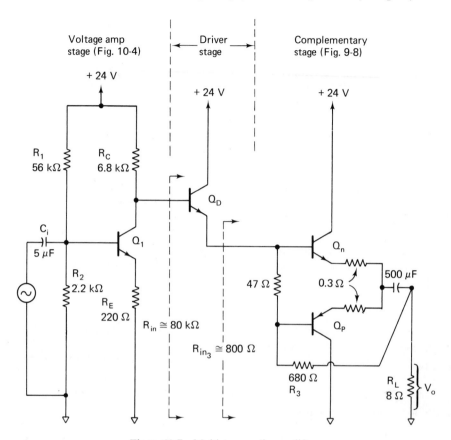

Figure 10-7 Multistage audio amplifier.

the complementary stage. For positive input signals Q_D and Q_n form a Darlington pair, while for negative input signals the Darlington pair is Q_D and Q_p.

Assume that $\beta = 100$ for each transistor in the audio amplifier of Fig. 10-7 and $R_L = 8\ \Omega$. Now, looking into the base of Q_D, the input resistance given by Eq. (10-7) is

$$\beta^2 R_L = (100)^2(8\ \Omega) = 80\ \text{k}\Omega$$

(*Note:* The input resistance of Q_D is the β of Q_D times either the β of Q_n or Q_p, not both. This is because when Q_n is on, Q_p is off, and when Q_p is on, Q_n is off.) Therefore, Q_D will present an input resistance of about 80 kΩ as the ac load on Q_1. R_{ac} (the ac load resistance) of Q_1 will equal $R_C \| 80$ kΩ or 6.8 k$\Omega \| 80$ k$\Omega \approx 6.8$ kΩ. Therefore, the voltage gain of Q_1 will remain $R_C/R_E \approx 30$. The voltage gain of emitter follower Q_D is about 1 and the voltage gain of $Q_n - Q_p$ is also about 1. Therefore, the voltage gain of the entire amplifier, V_o/E_i, is essentially established by Q_1 and will be slightly less than 30.

10-4.2 Biasing the Amplifier. As stated in Section 9-6.3, the emitters of Q_n and Q_p should be biased to $V_{CC}/2 = 12$ V. Idle current through R_3 will be 12 V/680 $\Omega = 17$ mA. This 17 mA is also the emitter current of Q_D, so Q_D will require a base-bias current of $I_E/(\beta + 1) = 17$ mA/101 ≈ 0.17 mA. This base-bias current is furnished by R_C and is negligible with respect to the almost 2-mA collector current of Q_1. If the collector of Q_1 is set at 13.2 V (with no signal) by adjusting R_1 or R_2, the emitters of Q_n and Q_p will be at 12 V. This is because the 0.6-V base–emitter voltages of both Q_D and Q_n clamp the emitters at 1.2 V below the collector voltage of Q_1.

10-4.3 Summary. Figure 10-7 illustrates an excellent low-cost audio amplifier. It also illustrates how and why BJTs are used to provide resistance multiplication. Input resistance of the amplifier is established mainly by R_2 at 2.2 kΩ.

10-5 Hybrid Power Amplifiers

10-5.1 Introduction. Hybrid power amplifiers are available that will control large amounts of power at a cost of 50 cents to $1 per watt. Two representative types, one domestic and one foreign are mentioned in the bibliography. The term *hybrid* means a circuit design that incorporates the best features of integrated circuits and discrete components. This technique results in a hermetically sealed package that may contain more than a dozen transistors, a half-dozen diodes, a dozen more resistors, and several small capacitors. The resulting package may occupy a volume equal to or smaller than a regular package of cigarettes and be able to deliver an rms power output of 100 W into a 4-Ω speaker load.

Typical hybrid power amplifiers have six to ten terminals. Two are used for an input signal, two for the output, and two for the power supply. Of the remaining terminals, some may be used to add capacitors that are too bulky for the package. These compensation capacitors may be required to prevent oscillations under certain operating conditions that will be defined by the manufacturer. Some of the remaining terminals may also be used to connect resistor(s) to set the voltage gain.

10-5.2 Operation. Figure 10-8 shows the schematic for a typical fixed-gain hybrid power amplifier. Voltage gain is determined by resistors located inside the hybrid module. Manufacturers do not specify voltage gain but

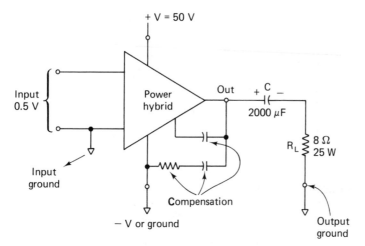

Figure 10-8 Hybrid power amplifier.

instead specify rms input voltage required (0.5 V) for full power output (25 W). Load R_L is also specified and is typically 8-Ω or 4-Ω speakers. Usually, maximum supply voltage is lowered for the lower resistance speakers to reduce load current and avoid burning out the hybrid power module. For example, a 50-V power supply could be used for 25 W to an 8-Ω load. But supply voltage should be reduced to about 35 V for a 4-Ω load.

The manufacturer must also furnish guidance on the size of heat sink required to operate the module at a particular room temperature. The module contains a metal base plate that serves as a heat sink and heat spreader. The base plate is bolted or epoxied to an aluminum sheet about $\frac{1}{16}$ inch thick. A typical sheet size required would be 4 inches by 6 inches to handle 25 W at room temperature. It is economical to bolt the module to the chassis (electrically insulated) and use the chassis as a heat sink.

10-5.3 Power Supply. Since no design work is required for the hybrid amplifier module and installation guidance is furnished by the manufacturer, we must also rely on the manufacturer to recommend the most economical supply. The manufacturer may also give helpful tips on wiring procedures to separate input and output ground currents. They should not share the same wire path. If they do, oscillations may result. An economical full-wave bridge rectifier is shown in Fig. 10-9. The ac secondary voltage of the transformer is rectified and converted to dc supply voltage by filter capacitor C. Proper grounding technique is shown to avoid ground loops.

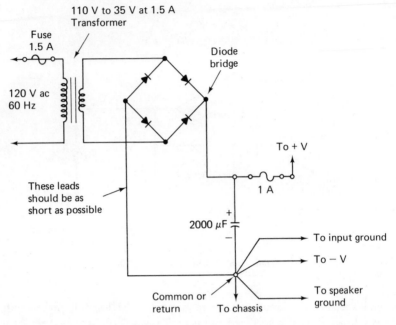

Figure 10-9 Power supply for Fig. 10-8.

Problems

10-1 Let R_E be changed to 125 Ω in Example 10-1. Also R_1 and R_2 are redesigned to set $I_C = 4$ mA and preserve $R_B = 2.5$ kΩ. Find the new (a) input resistance and (b) voltage gain.

10-2 Find the maximum possible output voltage for Problem 10-1.

10-3 If $R_{in2} = 5$ kΩ in Fig. 10-1, what is the gain of the Q_1 stage?

10-4 In Fig. 10-2, does adding R_{in2} raise or lower the voltage gain of Q_2?

10-5 In Fig. 10-1, is the maximum output swing of Q_1 increased or decreased when the Q_2 stage is added?

10-6 In Fig. 10-1, $E_i = 10$ mV, $V_{o1} = 75$ mV, and $V_o = 750$ mV. What is the gain of (a) Q_1; (b) Q_2; (c) the overall gain?

10-7 If $A_{v1} = 10$ and $A_{v2} = 20$ in a circuit similar to Fig. 10-1, find V_o when $E_i = 2$ mV.

10-8 A 4-Ω speaker is substituted for R_L in Fig. 10-3. Find V_o.

10-9 What is the combination of a common collector driving a common collector called?

10-10 If each transistor in Fig. 10-4 has a β of 50, calculate the current gain of the Darlington pair.

10-11 If the Darlington pair of Problem 10-10 has to supply an emitter current of 1.5 A, what is the base current?

10-12 What is the value of bias resistor R_B in Fig. 10-5 if $\beta = 50$ and $I_E = 1.5$ A. $R_E = 10\,\Omega$.

10-13 Calculate input resistance R_i if $\beta = 50$ and $R_E = 10\,\Omega$, for a circuit similar to Fig. 10-5(a).

10-14 Calculate input resistance R_{in} for Problem 10-13, $R_B = 63\,\text{k}\Omega$.

10-15 Consider Fig. 10-6(a), with $\beta = 50$, $R_E = 10\,\Omega$, and $R_i = 25\,\text{k}\Omega$. Calculate voltage gain A_V.

10-16 What is the input resistance looking into Q_D of Fig. 10-7 if $R_L = 4\,\Omega$.

10-17 For positive input signals into Fig. 10-7, which output transistor is on?

11

Transformer and Tuned Amplifiers

11-0 Introduction

In order to send many different signals simultaneously through the same air space or down the same wires, the signals must differ in frequency. It is the task of the tuned circuit to select one particular signal frequency range from all that may be present. Ideally, the tuned circuit selects only the desired frequencies and rejects all others. A transformer also does a job similar to a tuned circuit because the transformer transmits only ac signals and does not transmit direct-current signals.

Section 11-1 treats transformer-coupled amplifiers. This type of amplifier is capable of isolating one stage from another or an output stage from the load. In Section 11-2 the tuned amplifier is presented to show how a tuned circuit and amplifier are combined to provide both amplification and selection of a desired range. Section 11-3, on the tuned transformer amplifier, incorporates both the principles of transformers and parallel tuned circuits.

11-1 Transformer-Coupled Amplifiers

Figure 11-1 shows that load resistor R_L is inductively coupled to the collector of the transistor through the windings of a transformer. Remember that a transformer transforms only ac voltage and current, not dc voltage and current. Therefore, load resistor R_L has no dc current flowing through it. However, the transformer allows an ac output voltage V_o to be developed

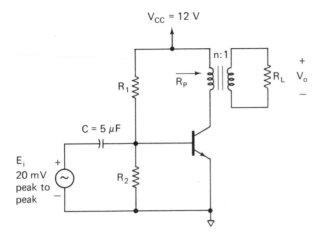

Figure 11-1 Transformer coupled circuit.

across R_L. To analyze this type of circuit, reflect R_L across the primary side of the transformer as a resistance R_p:

$$R_p = n^2 R_L \qquad (11\text{-}1)$$

R_p is the ac resistance appearing across the primary side, and n is the turns ratio from primary to secondary. If the number of turns on the primary side is greater than the number of turns on the secondary, n is greater than 1. Thus R_L may be a low-resistance speaker, but the transistor's collector will see a higher ac load resistance equal to R_p.

11-1.1 Graphical Analysis. Applying rules established in Chapter 6 for drawing a dc load line:

1. Horizontal axis intercept is at V_{CC}.
2. Vertical-axis intercept is at V_{CC}/R_{dc}, where R_{dc} is the dc resistance in the output loop. However, the dc primary resistance of the transformer is negligible, and there is no emitter resistor in Fig. 11-1. Then V_{CC}/R_{dc} = 12 V/0 Ω = ∞ (infinity). Therefore, the dc load line is a vertical line that intersects the V_{CE} axis at V_{CC}, as shown in Fig. 11-2.

The ac load line intersects the dc load line at operating point O. The operating point may be chosen and then set by R_1 and R_2. For maximum symmetrical output-voltage swing, the ac load line should intersect the V_{CE} axis at $2V_{CC}$. The current axis intercept for the ac load line is $2V_{CC}/R_p$, where R_p is given by Eq. (11-1).

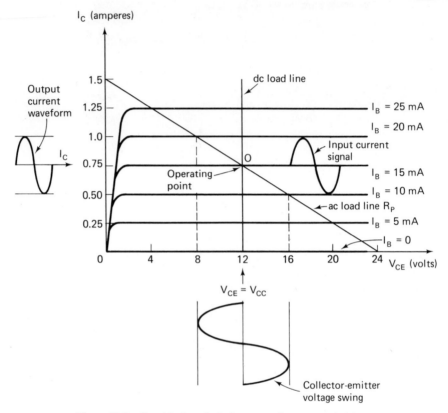

Figure 11-2 Graphical analysis for a transformer coupled load.

Example 11-1: In Fig. 11-1, $R_L = 4\,\Omega$ and $n = 2$. What are the horizontal and vertical axis intercepts for the ac load line?

Solution: From Eq. (11-1), $R_p = (2)^2\,4\,\Omega = 16\,\Omega$. The horizontal intercept is at $2(V_{CC}) = 2(12\text{ V}) = 24$ V. The vertical intercept is at $2V_{CC}/R_p = 24\text{ V}/16\,\Omega = 1.5$ A. Figure 11-2 shows both the dc and ac load line. The operating point is at $I_B = 15$ mA, $I_C = 0.75$ A, and $V_{CE} = 12$ V.

Example 11-2: What is the β of the transistor?
Solution: From Chapter 2,

$$\beta = \frac{I_C}{I_B} = \frac{0.75\text{ A}}{15\text{ mA}} = 50$$

The value of I_C and I_B is taken at the operating point, the intersection of the two load lines.

In the circuit of Fig. 11-1 the transistor must be capable of handling several amperes of collector current. Also note from the characteristics that the base current is in milliamperes. This transistor must be a power transistor. Characteristics and limitations of such devices will be studied in Chapter 16.

Example 11-3: E_i in Fig. 11-1 is set at 20 mV peak to peak, causing a 10-mA peak-to-peak base current signal, as shown in Fig. 11-2. What is the voltage gain of the circuit?
Solution: Figure 11-2 shows the peak-to-peak collector–emitter voltage swing of 8 V for a 10-mA and 20-mV peak-to-peak input signal.

$$A_V = \frac{16\text{ V} - 8.0\text{ V}}{20\text{ mV}} = \frac{8.0\text{ V}}{20\text{ mV}} = 400$$

A_V is the voltage gain from E_i to the collector of the transistor and *not* to R_L.

Example 11-4: What is the peak-to-peak voltage swing across R_L for Fig. 11-1? $n = 2$.
Solution: From Fig. 11-2 peak-to-peak collector–emitter voltage swing is 16 V − 8.0 V = 8.0 V:

$$V_{op-p} = \frac{1}{n} \times V_{CEp-p} = \tfrac{1}{2} \times 8.0\text{ V} = 4.0\text{ V}$$

For the transformer in Fig. 11-1, the number of turns on the secondary is less than the number of turns of the primary; therefore, the secondary voltage is stepped down. This transformer would be called a step-down transformer.

Example 11-5: R_L in Fig. 11-1 equals 4 Ω. What is (a) power developed across it and (b) current through it?
Solution:

(a) From Example 11-4, $V_{op-p} = 4.0$ V; converting to rms,

$$V_o = \frac{V_{op-p}}{2\sqrt{2}} = \frac{4.0\text{ V}}{2\sqrt{2}} \simeq 1.4\text{ V} \quad \text{and} \quad P_o = \frac{V_o^2}{R_L} = \frac{(1.4\text{ V})^2}{4\Omega} = 0.49\text{ W}$$

(b)

$$I_o = \frac{V_o}{R_L} = \frac{1.4\text{ V}}{4\Omega} = 0.35\text{ A}$$

From Fig. 11-2, the peak-to-peak collector–emitter current is 1.0 A − 0.50 A; converting to an rms value gives 0.5 A/$2\sqrt{2} \simeq 0.177$ A. [*Note:* The transformer has stepped down the load voltage by a ratio of $1/n$. The load current, I_o, is stepped up by the same ratio. $I_o = 2(0.177\text{ A}) \simeq 0.35$ A.]

11-2 Tuned Amplifiers

A resistor, inductor, and capacitor connected in parallel make up a *tuned circuit*. When this circuit is connected as the load on an amplifier, the resulting network is called a *tuned amplifier*, as shown in Fig. 11-3(a). This circuit is used to select and amplify only a narrow band of frequencies. Figure 11-3(b) is a plot of output voltage, V_o, versus frequency. This plot shows that there is one frequency at which V_o is a maximum. This frequency is called the *resonant frequency*, f_r, and its value is determined by

$$f_r = \frac{1}{2\pi\sqrt{LC}} \tag{11-2}$$

where L is the inductor value, in henries, and C is the value of the capacitor, in farads.

Example 11-6: What is the resonant frequency for Fig. 11-3(a) if $L = 1.0$ mH and $C = 0.01$ μF?
Solution: From Eq. (11-2),

$$f_r = \frac{1}{2(3.14)\sqrt{(1.0 \times 10^{-3})(0.01 \times 10^{-6})}} \simeq 50.4 \text{ kHz}$$

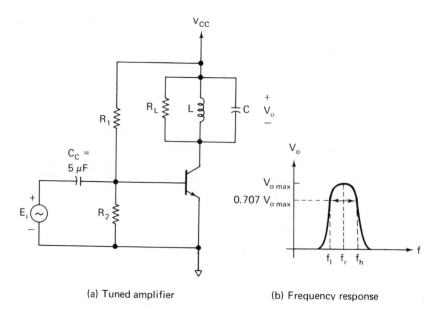

(a) Tuned amplifier   (b) Frequency response

Figure 11-3 Tuned amplifier and output-voltage frequency response.

11-2.1 Calculations at the Resonant Frequency. Figure 11-3(b) shows that output voltage, V_o, decreases for frequencies off the resonant frequencies. At f_r, inductive reactance X_L equals capacitive reactance, X_C, and the effective load is R_L, as shown in the hybrid-π model of Fig. 11-4. Output voltage, V_o, is a maximum at f_r and given by

$$V_{omax} = \beta I_b R_L \qquad (11\text{-}3)$$

Maximum output voltage may also be determined by graphical analysis, as will be illustrated in the next two examples.

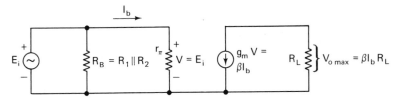

Figure 11-4 Hybrid-π model for Fig. 13-3(a) at f_r.

Example 11-7: The I_C-V_{CE} characteristics for the transistor in Fig. 11-3(a) are shown in Fig. 11-5. If $R_L = 4$ kΩ, $V_{CC} = 12$ V, and the ac input peak-to-peak base current is 40 μA, determine the output voltage.

Solution: For dc conditions the inductor appears to be a short and the dc resistance in the output loop is zero. The dc load is a vertical line that intersects the horizontal axis at $V_{CC} = 12$ V, as shown in Fig. 11-5. Choosing an operating point midway on the load line locates point O at $I_B = 30$ μA, $I_C = 3$ mA, and $V_{CE} = 12$ V. A peak-to-peak base-current signal of 40 μA (or 20-μA peak) varies the operating point between $I_B = 30$ μA $+ 20$ μA $= 50$ μA and $I_B = 30$ μA $- 20$ μA $= 10$ μA. The corresponding peak values of output voltage are 4 V at $I_B = 50$ μA and 20 V at $I_B = 10$ μA, as shown in Fig. 11-5. Therefore, the peak-to-peak output voltage swing is 20 V $- 4$ V $= 16$ V.

Example 11-8: Use the values given in Example 11-7 to determine ac output voltage using Eq. (11-3).

Solution: Since R_L and V_{CC} are the same as in Example 11-7, the load is the same, and choosing the same operating point we may calculate β:

$$\beta = \frac{I_C}{I_B} = \frac{3\text{mA}}{30 \ \mu\text{A}} = 100$$

From Eq. (11-3), at f_r, $V_o = 100(40 \ \mu\text{A})(4 \text{ k}\Omega) = 16$ V. This checks with the answer of Example 11-7.

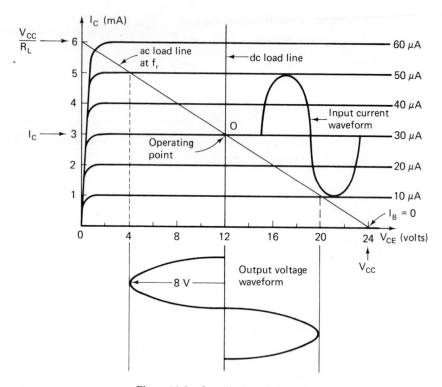

Figure 11-5 Graphical analysis at f_r.

In addition to the output voltage being a maximum at f_r, the voltage gain, A_V, is also a maximum, and, for Fig. 11-3(a),

$$A_V = \frac{V_o}{E_i} = \frac{\beta R_L}{r_\pi} \qquad (11\text{-}4)$$

Example 11-9: Calculate the voltage gain for Fig. 11-3(a) using the operating point shown in Fig. 11-5.

Solution: From Fig. 11-5, $I_C = 3$ mA, and from Eq. (7-5a),

$$r_\pi = \frac{\beta \times 30}{I_C(\text{mA})} = \frac{100 \times 30}{3} = 1\text{k}\Omega$$

The value of β was calculated in Example 11-8. Applying Eq. (11-4),

$$A_V = \frac{100(4\text{ k}\Omega)}{1\text{ k}\Omega} = 400$$

Example 11-10: What is the value of E_i in Fig. 11-3(a) to develop an output voltage at f_r of 16 V peak to peak?

Solution: Since $A_V = V_o/E_i$ and $A_V = 400$ from Example 11-9, $E_i = V_o/A_V = 16$ V/400 $= 40$ mV. (*Note:* In Fig. 11-4, $E_i = I_b r_\pi$ from Example 11-7, $I_b = 40$ μA peak to peak, and from Example 11-9 $r_\pi = 1$ kΩ. Then $E_i = (40$ μA)(1 kΩ) $= 40$ mV.)

11-2.2 Calculations at f_l and f_h. Figure 11-3(b) shows that as frequency is varied from f_r, there are two frequencies (one above f_r and one below f_r) at which $V_o = 0.707V_{omax}$. These frequencies are designated f_l and f_h. f_l is called *lower cutoff frequency* and f_h is called *upper cutoff frequency*. The band of frequencies between f_h and f_l is called *bandwidth*, BW. A tuned amplifier is classified as a narrow-band tuned amplifier if the bandwidth is equal to or less than one-tenth of the resonant frequency.

$$BW \leq 0.1 f_r \qquad \text{narrow-band tuned amplifier} \qquad (11\text{-}5)$$

We shall limit our discussion to narrow-band tuned amplifiers.

At both f_l and f_h, the magnitude of the ac load is

$$|Z| = 0.707 R_L \qquad \text{at } f_l \text{ and } f_h \qquad (11\text{-}6)$$

and the output voltage at these frequencies is

$$V_o = 0.707 V_{omax} \qquad \text{at } f_l \text{ and } f_h \qquad (11\text{-}7)$$

as shown in Fig. 11-3(b).

At the resonant frequency the ac load is R_L, but at f_l and f_h, the ac load is given by Eq. (11-6). This means that there is a new ac load line, which may be constructed by knowing the operating point and the horizontal axis intercept. This intercept is calculated by

$$V_{CE} \text{ (intercept)} = V_{CE} + I_C |Z| \qquad (11\text{-}8)$$

where V_{CE} is dc operating voltage, I_C is dc operating-point current, and $|Z|$ is given by Eq. (11-6).

Example 11-11: Use the values of Example 11-7 to draw (a) the dc load line (b) the ac load line and to determine (c) peak-to-peak output voltage.
Solution: Horizontal axis intercept is $V_{CE} = V_{CC} = 12$ V and vertical axis intercept is $V_{CC}/R_{dc} = 12$ V/0 $\Omega = \infty$. Thus the dc load line is a vertical line. Dc resistance in the collector output loop is zero because at dc (frequency equals zero) inductive reactance, $X_L = 2\pi f L$, equals zero. From Example 11-7, (b) the dc operating point is $I_B = 30$ μA, $I_C = 3$ mA, and $V_{CE} = 12$ V. Applying Eq. (11-8),

$$V_{CE} \text{ (intercept)} = 12 \text{ V} + (3 \text{ mA})(0.707)(4 \text{ k}\Omega) \simeq 12 \text{ V} + 8 \text{ V} = 20 \text{ V}$$

Figure 11-6 shows the dc load line, the ac load line at f_l and f_h, and, for a

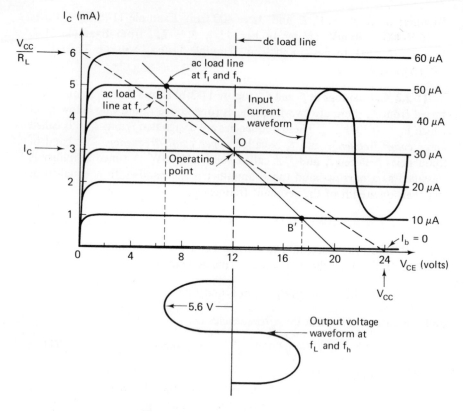

Figure 11-6 Ac load lines at f_l and f_h.

comparison, the ac load line at f_r. (c) The 40 μA peak-to-peak input base current signal is also drawn in Fig. 11-6.

When the input current goes to 30 μA + 20 μA = 50 μA, the collector–emitter voltage is 5.3 V at *B*. Then when I_B equals 30 μA − 20 μA = 10 μA, $V_{CE} \simeq 16.6$ V at *B'*. Therefore, peak-to-peak output voltage is 16.3 − 5.0 V = 11.3 V. Comparing this answer to that obtained using Eq. (11-5),

$$V_o = 0.707 V_{omax} = (0.707)(16 \text{ V}) = 11.3 \text{ V}$$

V_{omax} was obtained from Example 11-7.

Figure 11-7 shows the ac equivalent circuit for frequencies f_l and f_h, and the output voltage magnitude may also be determined by

$$V_o = \beta I_b(0.707 R_L) \tag{11-9}$$

Example 11-12: Calculate V_o peak to peak using Eq. (11-9) for the circuit values given in Example 11-7.

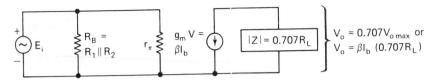

Figure 11-7 Ac equivalent circuit at f_l and f_h.

Solution: From Example 11-8, $\beta = 100$. Applying Eq. (11-9),

$$V_o = 100(40 \ \mu A)(0.707)(4 \ k\Omega) = 11.3 \ V$$

This answer checks with those calculated in Example 11-11.

Figure 11-6 shows that the ac load line is different for all frequencies. As the frequency is varied from f_r (either increasing or decreasing), the ac load pivoted at the operating point and rotates in a clockwise direction, as shown in Fig. 11-8. The limit is when the ac load line reaches the vertical position (dc load line), where the output voltage is zero.

11-2.3 Bandwidth. The bandwith is calculated from the difference between f_h and f_l:

$$BW = f_h - f_l \tag{11-10}$$

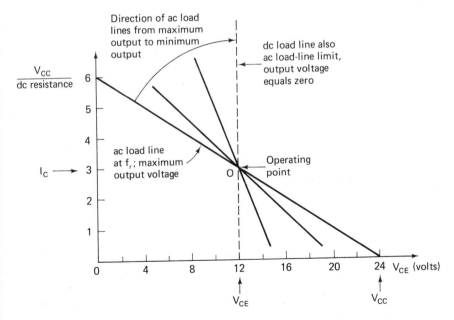

Figure 11-8 Direction of ac load lines for various frequencies of f_r.

When testing a tuned amplifier, keep E_i constant and increase the frequency until V_o drops to $0.707V_{omax}$. Record the frequency as f_h. Now decrease the frequency until V_o again drops to $0.707V_{omax}$, record the frequency as f_l, use Eq. (11-10), and calculate bandwidth. Bandwidth may also be calculated from circuit quantities:

$$BW = \frac{1}{2\pi RC} \tag{11-11}$$

Example 11-13: For the circuit values used throughout this section—$R_L = 4\,k\Omega$, $L = 1.0\,mH$, and $C = 0.01\,\mu F$—calculate the bandwidth for Fig. 11-3(a).
Solution: From Eq. (11-11),

$$BW = \frac{1}{2(3.14)(4 \times 10^3)(0.01 \times 10^{-6})} = 3.98\ kHz$$

[*Note:* Bandwidth depends on the load resistance but resonant frequency does not; see Eq. (11-2).]

11-2.4 Quality Factor. A quality factor, Q, is used to compare tuned amplifiers to see which one is more selective. That is, which amplifier has a smaller bandwidth at the resonant frequency?

$$Q = \frac{fr}{BW} \tag{11-12}$$

Quality factor Q may also be determined from circuit values,

$$Q = \frac{R}{2\pi f_r L} \tag{11-13a}$$

or

$$Q = 2\pi f_r\, RC \tag{11-13b}$$

Example 11-14: Use the circuit values in Example 11-6 to show that Eqs. (11-12) and (11-13) produce the same result. $R_L = 4\,k\Omega$.
Solution: From Example 11-6, $f_r = 50.4\,kH$, and from Example 11-13, $BW = 3.98\,kHz$; applying Eq. (11-12),

$$Q = \frac{50.4\ kHz}{3.98\ kHz} \simeq 12.7$$

Using Eq. (11-13b),

$$Q = (6.28)(50.4 \times 10^3)(4 \times 10^3)(0.01 \times 10^{-6}) \simeq 12.7$$

11-3 Tuned Transformer Amplifier

The tuned transformer amplifier of Fig. 11-9 uses the principle of the transformer-coupled load developed in Section 11-1 and the tuned circuit developed in Section 11-2. The easiest and best way to analyze this circuit is to reflect R_L from the secondary to the primary of the transformer and treat the problem as those in Section 11-2. The advantage of using the transformer is that the resistance and/or capacitance may be reflected from the secondary to the primary to give the desired values for a particular resonant frequency and bandwidth. By reflecting either resistance and/or capacitance, we may be able to use stock items instead of special-order items. Also, we obtain isolation between the collector of the transistor and the load.

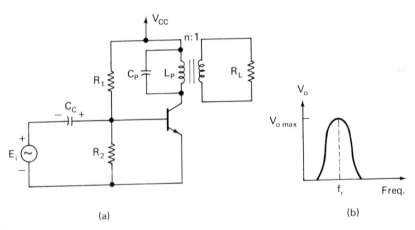

Figure 11-9 (a) Tuned transformer amplifier; (b) output-voltage wave form.

Example 11-15: For Fig. 11-9, let $R_L = 6 \text{ k}\Omega$, $C_p = 40 \text{ pF}$, $L_p = 100 \text{ mH}$, and $n = 10$. Calculate (a) resonant frequency; (b) bandwidth; (c) Q of the circuit.
Solution:

(a) From Eq. (11-2),

$$f_r = \frac{1}{2\pi(\sqrt{100 \times 10^{-3}})(40 \times 10^{-12})} \simeq 79.6 \text{ kHz}$$

(b) Reflecting R_L onto the primary side of the transformer and using Eq. (11-1),

$$R_p = (10)^2(6 \text{ k}\Omega) = 600 \text{ k}\Omega$$

Applying Eq. (11-11),

$$BW = \frac{1}{2\pi(600 \times 10^3)(40 \times 10^{-12})} \simeq 6.63 \text{ kHz}$$

Note that this is a narrow-band tuned amplifier because $BW \leq 0.1f_r$.
(c) From Eq. (11-12),

$$Q = \frac{79.6 \text{ kHz}}{6.63 \text{ kHz}} = 12$$

As a check, use Eq. (11-13a) or (11-13b); applying Eq. (11-13b),

$$Q = 2\pi(79.6 \times 10^3)(600 \times 10^3)(40 \times 10^{-12}) = 12$$

The value of R used in Eqs. (11-1) and (11-13) is the resistance on the primary side of the transformer, R_p, and not R_L.

Problems

11-1 Change n to 3.47 in Example 11-1. What are the horizontal and vertical axis intercepts for the ac load line? Locate the operating point.

11-2 $R_p = 48 \, \Omega$, $I_B = 5 \text{ mA}$, and peak-to-peak base-current signal is 5 mA in Fig. 11-2. Sketch the collector–emitter voltage swing.

11-3 If $n = 3.47$ and peak-to-peak collector–emitter voltage swing is 12 V in Fig. 11-1, find $V_{op\text{-}p}$.

11-4 In Fig. 11-1, $V_{op\text{-}p} = 10 \text{ V}$ and $R_L = 5 \, \Omega$. Find the power developed.

11-5 If $R_L = 10 \, \Omega$ in Fig. 11-1 and R_p should be 1000 Ω, what transformer turns ratio is required?

11-6 $C = 0.04 \, \mu\text{F}$ and $L = 100 \, \mu\text{H}$ in Fig. 11-3. Find f_r.

11-7 Given $\beta = 100$, $R_L = 4.7 \text{ k}\Omega$ and peak-to-peak base-current swing is 10 μA in Fig. 11-3. Find $V_{op\text{-}p}$ at the resonant frequency.

11-8 Find the voltage gain for Problem 11-7 if $I_C = 0.5 \text{ mA}$.

11-9 Find V_o at f_l and f_h if $V_{omax} = 10 \text{ V}$ at f_r.

11-10 What is the bandwidth for a tuned amplifier when $f_h = 41 \text{ kHz}$ and $f_l = 38 \text{ kHz}$?

11-11 In Fig. 11-3, $R_L = 5 \text{ k}\Omega$, $V_{CC} = 10 \text{ V}$, $I_C = 2.0 \text{ mA}$, and $I_B = 20 \, \mu\text{A}$. Draw the ac load lines at f_r, f_l, and f_h using the characteristic curves of Fig. 11-5.

11-12 Sketch output-voltage waves for $I_{bp\text{-}p} = 20 \, \mu\text{A}$ at both f_r and f_l in Problem 11-11.

11-13 In Fig. 11-3, $C = 0.04\ \mu\text{F}$ and $R = 10\ \text{k}\Omega$. Find the bandwidth.

11-14 In Fig. 11-3, $C = 250\ \text{pF}$, $L = 1\ \text{mH}$, and $R_L = 5\ \text{k}\Omega$. Find (a) f_r; (b) BW.

11-15 Find Q in Problem 11-14.

11-16 A circuit has $Q = 20$ at $f_r = 1\ \text{MHz}$. Find the bandwidth.

11-17 C is changed to 4000 pF in Example 11-15. Find (a) f_r; (b) BW; (c) Q.

12

Junction Field-Effect Transistors

12-0 Introduction

Until now we have discussed and studied bipolar junction transistors (BJTs). The word *bipolar* is used because these devices depend on the motion of two charge carriers, holes and electrons. The operation of field-effect transistors (FETs) depends only on the motion of one charge carrier, either holes or electrons, but not both. For this reason, FETs are classified as unipolar junction transistors.

There are two types of field-effect transistor: (1) the *junction field-effect transistor* (JFET), and (2) the *insulated-gate field-effect transistor* (IGFET), more commonly known as the *metal-oxide-semiconductor field-effect transistor* (MOSFET). This chapter treats only JFETs; Chapter 13 deals with MOSFETs.

Advantages of FETs over BJTs are: (1) high input impedance, (2) lower noise, (3) fewer steps in the manufacturing process, (4) more devices can be packaged into a smaller area for integrated circuit arrays, and (5) no thermal runaway. Disadvantages of FETs are: (1) poor high-frequency performance, (2) low power-handling ability, and (3) low values of voltage gain.

12-1 Junction Field-Effect Transistors

Figure 12-1 illustrates a cross section of the two types of JFETs. Figure 12-1(a) is an *n*-channel JFET, Fig. 12-1(c) that of a *p*-channel JFET. The *n*-channel JFETs are manufactured by diffusing *p*-type semiconductor material

190

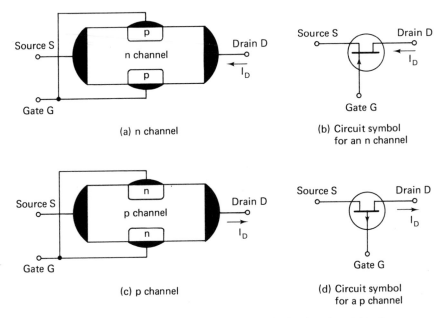

Figure 12-1 Construction and circuit symbols for junction field-effect transistors.

into a bar of *n*-type semiconductor material. The *p*-channel JFETs are made by diffusing *n*-type material into a bar of *p*-type material. Thus JFETs are classified according to the material of the channel and not the diffusing material. This is because the channel type tells whether free electrons in the *n*-channel or free holes in the *p*-channel can conduct channel current.

Ohmic contacts are then connected to both ends of the channel and to the diffused material. Connections to the channel are called *source, S,* and *drain, D.* The direction of conventional drain current through an *n*-channel JFET is from drain to source and is designated I_D. Connections from the diffused material end at the *gate,* G, terminal. Circuit symbols for an *n*-channel and *p*-channel JFET are shown in Fig. 12-1(b) and (d), respectively.

12-1.1 JFET Operation. Operation of JFETs first requires making the drain terminal of proper polarity with respect to the source terminal, to establish drain current, I_D. The value of I_D depends on the width of the channel. If the width of the channel is decreased, I_D decreases. If the channel width increases, I_D increases. The width of the channel is controlled by reverse-biasing the *pn* junction between gate and source. Controlling the reverse bias at the junction controls the space-charge region, which in turn controls the width of the channel.

In Fig. 12-2, V_{DD} establishes the proper polarity from drain to source

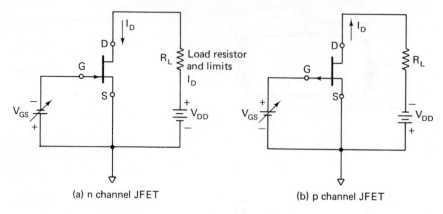

(a) n channel JFET   (b) p channel JFET

Figure 12-2   Biasing for both types of JFETs.

and R_L limits the maximum drain current that can flow. V_{GS} reverse-biases the *pn* junction between gate and source. Figure 12-2(a) shows the proper polarity of V_{DD} and an *n*-channel JFET to attract electrons. Figure 12-2(b) shows the proper polarity of V_{DD} to attract holes for a *p*-channel JFET.

12-2 Current-Voltage Characteristics of a JFET

The current–voltage characteristics of a JFET are a plot of drain current, I_D, versus drain-to-source voltage, V_{DS}, for different values of gate-to-source voltage, V_{GS}. To measure the I_D–V_{DS} characteristics of an *n*-channel JFET, begin by short-circuiting gate to source to ensure that gate voltage V_{GS} equals zero. Next increase V_{DD} to make the drain more positive than the source. When V_{DS} increases from 0 to about 1 V, I_D also increases, exactly like the I–V characteristics of a resistor. This is shown in Fig. 12-3 from *O* to *A*. The *O–A* portion of the I_D–V_{DS} characteristic is called the *ohmic* region.

As V_{DS} is increased beyond 4 V, I_D remains relatively constant. Operation along this flat portion of the characteristics is referred to as (1) *pentode region* (this term is carried over from vacuum tubes), (2) *saturation region*, to indicate that the JFET is carrying all possible current, or (3) *constant-current region*, to indicate that the drain current remains constant as V_{DS} increases. The value of current that flows in the saturation region when $V_{GS} = 0$ is called the *drain-to-source saturation current*, I_{DSS}.

12-2.1 On Resistance. The ratio of $\Delta V_{DS}/\Delta I_D$ at the origin, where $V_{GS} = 0$, is called the drain or *on-resistance*, $r_{ds(on)}$. In equation form,

$$r_{ds(on)} = \frac{\Delta V_{DS}}{\Delta I_D} \qquad \text{for } V_{GS} = 0 \text{ V} \qquad (12\text{-}1)$$

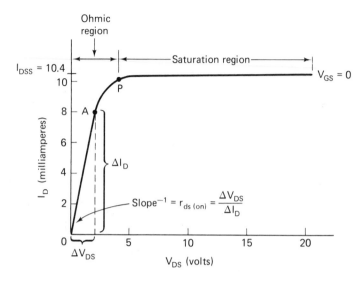

Figure 12-3 I_D-V_{DS} characteristic curve for $V_{GS} = 0$.

From Fig. 12-3,

$$r_{ds(on)} \approx \frac{2\,V}{8\,mA} = 250\,\Omega$$

Values of $r_{ds(on)}$ range from $100\,\Omega$ to several thousand ohms. $r_{ds(on)}$ for a JFET is similar to the collector-to-emitter saturation resistance of a bipolar junction transistor. R_{sat} of BJTs is usually in the neighborhood of a few ohms and therefore is much lower than $r_{ds(on)}$ of JFETs.

12-2.2. Saturation Region. Figure 12-4 is used to explain why the drain current remains constant for values of drain-to-source voltage greater than 4 V. Assume that $V_{DS} = 4$ V. This voltage divides equally between drain and source because the semiconductor material from drain to source is uniform and there are no *pn* junctions to cross. As shown in Fig. 12-4, the source ends of both *pn* junctions are reverse-biased by 2 V. The drain ends of both *pn* junctions are reverse-biased by 3 V. The width of a space-charge region is directly proportional to the amount of the reverse bias. Since the reverse-bias voltage is greater at the drain end, the space charge is wider, and therefore the width of the channel is narrower. As V_{DS} is increased, tending to increase drain current, the channel is narrowed to offset the increase in drain current and hold it constant.

12-2.3 Pinchoff Voltage. The voltage that causes the space-charge regions to touch and close the channel is called *pinchoff voltage*, V_p. Pinchoff voltage

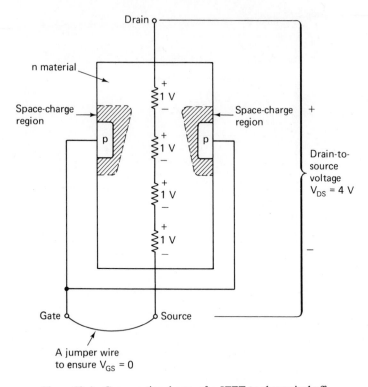

Figure 12-4 Cross-sectional area of a JFET to show pinchoff.

depends on both drain-to-source voltage, V_{DS}, and gate-to-source voltage V_{GS}, or

$$V_p = V_{DS} + V_{GS} \qquad (12\text{-}2)$$

When $V_{GS} = 0$, $V_p = V_{DS}$. Pinchoff voltage can be estimated from the I_D–V_{DS} characteristic for $V_{GS} = 0$ by finding the center of the transition curve between ohmic region and saturation region. For point P on the FET characteristic curve of Fig. 12-3, at $V_{GS} = 0$, $V_p = V_{DS} = 4$ V.

12-2.4 Family of Characteristic Curves. In Fig. 12-2, set $V_{GS} = -1$ V (-1 volt because all voltages are taken with respect to the source) to help in reverse-biasing the pn junction. Now when V_{DS} is increased, less drain current flows for the same value of V_{DS} when compared with $V_{GS} = 0$. This is because the channel is pinched off at a lower value of V_{DS}. In accordance with Eq. (11-2),

$$V_{DS} = V_p - V_{GS} = 4\,\text{V} - 1\,\text{V} = 3\,\text{V}$$

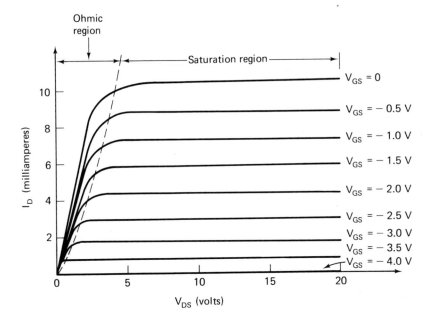

Figure 12-5 Characteristic curves for a JFET.

If V_{GS} is increased from 0 to -4 V in steps of -0.5 V and all the curves plotted on the same graph, the result is Fig. 12-5.

12-3 Biasing

12-3.1 Source-Bias Resistor. The disadvantage with Fig. 12-2 is the need for two dc power supplies to operate the JFET. Supply V_{GS} may be removed and replaced by two 10% resistors R_G and R_S to bias the JFET, as shown in Fig. 12-6(a). This design change produces considerable savings in the overall size, weight, and cost of a package using a JFET.

Gate resistor R_G is in the circuit to complete a closed path from gate to source. Current through R_G is negligible because it is the same current that flows through the reverse-biased gate-to-source junction. Values of this gate-to-source current are in the order of nanoamperes. R_G must not be small, however, because we wish to preserve the characteristic of high input resistance, which is the reason for using a JFET in the first place. Typical values for R_G are between 1 and 10 MΩ. Although R_G is large, the current through it is so small that for all practical purposes the voltage drop across R_G may be neglected.

Therefore, the voltage V_{GS} that reverse-biases the gate-to-source junction is set by I_D and R_s.

$$|V_{GS}| = I_D R_s \qquad (12\text{-}3)$$

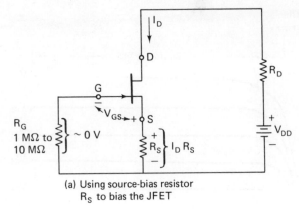

(a) Using source-bias resistor
R_S to bias the JFET

$V_{GS} = 0, R_S = 0$

$V_{GS} = -0.5$ V, $R_S \cong 57 \ \Omega$

$V_{GS} = -1.0$ V, $R_S \cong 137 \ \Omega$

$V_{GS} = -1.5$ V, $R_S \cong 254 \ \Omega$

$V_{GS} = -2.0$ V, $R_S \cong 455 \ \Omega$

$V_{GS} = -2.5$ V, $R_S \cong 833 \ \Omega$

$V_{GS} = -3.0$ V, $R_S \cong 1666 \ \Omega$
$V_{GS} = -3.5$ V, $R_S \cong 3889 \ \Omega$

$V_{GS} = -4.0$ V, $R_S \cong \infty$

(b) Characteristic curves using the circuit in (a)

Figure 12-6 Biasing a JFET using the source-bias resistor R_s.

By varying R_s we can obtain the same family of characteristic curves as given in Fig. 12-5. The values shown in Table 12-1 were obtained by use of Eq.

Table 12-1 Characteristic curves of a JFET

$|V_{GS}|$, V	I_D, mA	$R_s = V_{GS}/I_D$, Ω
0	10.5	0
0.5	8.8	57
1.0	7.3	137
1.5	5.9	254
2.0	4.4	455
2.5	3.0	833
3.0	1.8	1666
3.5	0.8	3889
4.0	0	∞

(12-3) and Fig. 12-5. Figure 12-6(b) shows the characteristic curves with the value of R_s needed in Fig. 12-6(a) to give the same curve as the dc gate supply V_{GS} in Fig. 12-2.

12-3.2 Load Line. Writing a loop equation around the output loop in Fig. 12-7(a) yields

$$V_{DD} = V_{DS} + I_D(R_L + R_s) \qquad (12\text{-}4)$$

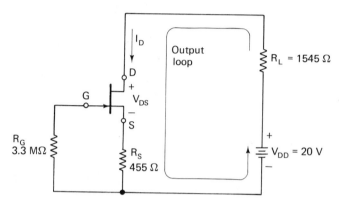

(a) Biasing for an n channel JFET

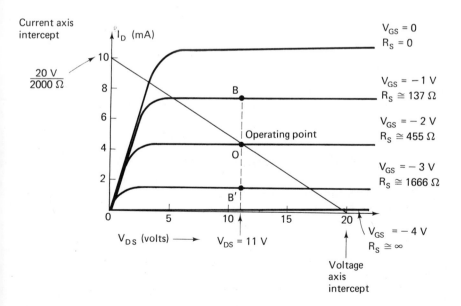

Figure 12-7 Load-line analysis for JFETs.

To obtain graphical solutions to JFET problems we need to plot Eq. (12-4) on the *I–V* characteristics. Equation (12-4) is an equation of a straight line and only two points are needed to plot it accurately. The easiest two points to obtain are the vertical or current axis and horizontal or voltage axis intercepts.

Vertical axis intercept: When the JFET acts as a *short circuit*, that is, when $V_{DS} = 0$, Eq. (12-4) reduces to

$$V_{DD} = 0 + I_D(R_L + R_s)$$

or
$$I_D = \frac{V_{DD}}{R_L + R_s} \qquad (12\text{-}5a)$$

Using the circuit values in Fig. 12-7(a),

$$I_D = \frac{20\ \text{V}}{(1545 + 455)\ \Omega} = 10\ \text{mA} \qquad \text{at } V_{DS} = 0\ \text{V}$$

This point is shown in Fig. 12-7(b) as the *current axis intercept*.

Horizontal axis intercept: When the JFET acts as an *open circuit*, that is, when $I_D = 0$, Eq. (12-4) reduces to

$$V_{DD} = V_{DS} \qquad \text{at } I_D = 0 \qquad (12\text{-}5b)$$

For the circuit of Fig. 12-7(a), the *voltage axis intercept* is 20 V and shown in Fig. 12-7(b).

Keep in mind that the current and voltage intercepts are only the end points, and that under normal operating conditions the JFET must have an operating point located on the load line between the two intercepts.

For the circuit of Fig. 12-7(a) with a source resistance $R_s = 455\ \Omega$ the operating point will be at the intersection of the load line and $V_{GS} = -2$-V curve, as shown in Fig. 12-7(b).

12-4 Common-Source Amplifier

The biasing arrangement of Fig. 12-8(a) is the same as that of Fig. 12-7(a). Drain current I_D produces a dc voltage across R_s equal to a value of V_{GS}, and a dc voltage from drain-to-source V_{DS} that ensures the operating point is located in the saturation region. With biasing established, we are now ready to apply the ac signal E_i. Set the frequency of E_i at 1 kHz so that the capacitive reactance ($X_C = 1/2\pi f C$) of C_c and C_s is negligible with respect to circuit resistors. Therefore, the coupling capacitor C_c and bypass capacitor C_s are ac short circuits. With this in mind and the fact that supply voltage

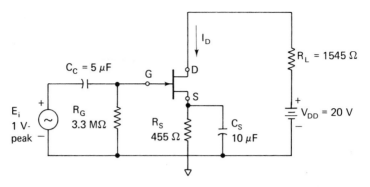

(a) Common-source amplifier

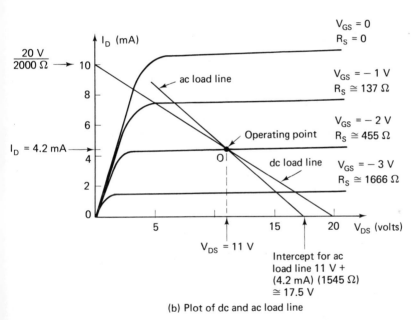

(b) Plot of dc and ac load line

Figure 12-8 Common-source amplifier and characteristics.

V_{DD} also has negligible ac resistance, we are ready to draw an ac load line along with the dc load line on the characteristic curves of Fig. 12-7(b).

12-4.1 AC Load Line. The operating point is located by the intersection of the dc load line with the V_{GS} curve that corresponds to $R_s = 455\,\Omega$. If the ac resistance in the output loop is different from the dc resistance, the ac load line is different from the dc load line. However, the ac load line must also go through the operating point. Thus we know one point on the ac load line. If we are able to determine another point, the ac load line can be quickly constructed. The horizontal intercept for an ac load line is

$$V_{DS} + I_D R_{ac} \tag{12-6}$$

where V_{DS} is the drain-to-source operating point voltage, I_D is the drain current at the operating point, and R_{ac} is the ac resistance in the output loop.

Example 12-1: Draw the dc and ac load line for Fig. 12-8(a).
Solution: The dc load line is the same as that of Fig. 12-7(b) and is calculated from
  Horizontal axis intercept:

$$V_{DS} = V_{DD} = 20 \text{ V}$$

Vertical axis intercept:

$$I_D = \frac{V_{DD}}{R_L + R_s} = \frac{20 \text{ V}}{1545 \, \Omega + 455 \, \Omega} = 10 \text{ mA}$$

The dc load line is redrawn on Fig. 12-8(b). The operating point is located at $V_{GS} = 2$ V, which is also one point on the ac load line. From Eq. (12-6) the ac horizontal axis intercept is 11 V + (4.2 mA)(1545 Ω) $\approx$ 17.5 V. Figure 12-8(b) shows both the dc and ac load lines.

12-4.2 Graphical Analysis. As with bipolar junction transistors, graphical analysis for field-effect transistor circuits is a visual aid to better understand how output current and voltage vary as input voltage changes. In Fig. 12-8(a), E_i is set at 1 V peak. Since R_s is bypassed by C_s, the ac resistance from source to ground is negligible and all of E_i appears between gate and source to vary the dc bias voltage V_{GS}. At the operating point in Fig. 12-8(b), the dc bias voltage, V_{GS}, equals -2 V. As E_i goes positive, the *pn* junction between gate and source becomes less reverse-biased ($V_{GS} + E_i = -2$ V + 1 V = -1 V), which results in an increase in drain current. When E_i goes negative the *pn* junction is reverse-biased by a larger value of voltage ($V_{GS} - E_i = -2$ V $- 1$ V = -3 V), which causes less drain current and an increase in drain-to-source voltage. Figure 12-9 shows how the drain current and drain-to-source voltage varies as input voltage E_i varies by 1 V peak. Note that the operating path *BOB'* lies along the ac load line.

Voltage gain, A_v, is the ratio of output voltage divided by input voltage. From Fig. 12-9 the ratio of peak-to-peak output voltage divided by peak-to-peak input voltage is

$$A_v = \frac{\Delta V_o}{\Delta E_i} \approx \frac{15 \text{ V} - 7 \text{ V}}{3 \text{ V} - 1 \text{ V}} = 4$$

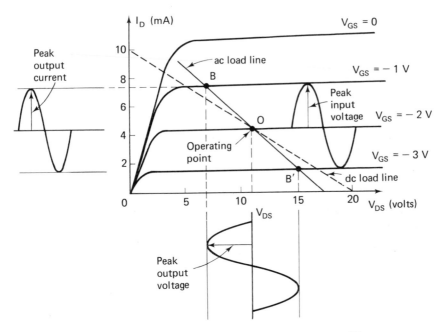

Figure 12-9 Graphical analysis of a common-source amplifier.

As was stated in the introduction, voltage gains for JFET circuits are usually lower than that for BJTs.

12-5 AC Model and Parameters for a JFET

12-5.1 Forward Transconductance. For both bipolar and field-effect transistors, we are interested in an output–input relation. This number may be used as one factor to compare devices. For BJTs connected in the common-emitter configuration, this factor is β and found by Eq. (4-2). For field-effect transistors, the output–input relation is the ratio of drain current, I_D, to gate-to-source voltage, V_{GS}, for a constant value of drain-to-source voltage, V_{DS}. Since an ac voltage will be applied to the JFET, we should determine the ratio of ΔI_D (change in I_D) to ΔV_{GS} (change in V_{GS}). This ratio is called *forward transconductance*, g_{fs}:

$$g_{fs} = \frac{\Delta I_D}{\Delta V_{GS}} \qquad V_{DS} = \text{constant} \qquad (12\text{-}7)$$

Example 12-2: Calculate g_{fs} at the operating point in Fig. 12-7(b).

Solution: Taking a value for ΔV_{GS} that changes $1V$ above and $1V$ below the operating point yields

$$\text{at } V_{DS} = 11 \text{ V}: \quad g_{fs} = \frac{7.5 \text{ mA} - 1.6 \text{ mA}}{3 \text{ V} - 1 \text{ V}} = 2.95 \times 10^{-3} \text{ } \mho$$

$$= 2950 \text{ } \mu\mho$$

When characteristic curves are not available, values for forward transconductance may be obtained from a manufacturer's data sheet. Typical values are from 1000 to 5000 $\mu\mho$. Some manufacturer's and texts use g_m or y_{fs} in place of g_{fs}. In this text we shall use g_{fs} exclusively.

12-5.2 AC Model. Figure 12-10(b) is an ac model of Fig. 12-10(a) operating at 1 kHz. The *pn* junction between gate and source is reverse-biased and for all practical purposes conducts zero current. Therefore, the model shows an open circuit between gate and source. The ac voltage V causes an

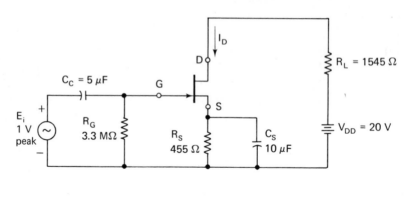

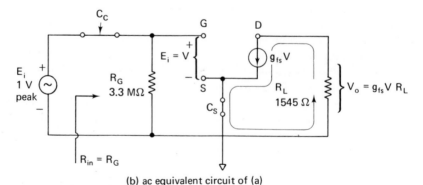

(b) ac equivalent circuit of (a)

Figure 12-10 Complete circuit and ac model for a common-source amplifier.

increase and decrease in the voltage between gate and source. This, in turn, causes a change in the width of the channel that varies the drain current. If the drain current is varying sinusoidally, an ac voltage is produced across R_L. Ac drain current is shown as g_{fs} V.

12-5.3 Voltage Gain. In Fig. 12-10(b) ac output voltage V_o is the product of g_{fs} V and R_L. Since the ac gate-to-source voltage V is approximately E_i, we may express V_o as

$$V_o = g_{fs}VR_L = g_{fs}E_iR_L \qquad (12\text{-}8a)$$

and voltage gain is

$$A_v = \frac{V_o}{E_i} = g_{fs}R_L \qquad (12\text{-}8b)$$

where g_{fs} is found by using Eq. (12-7).

Example 12-3: Use Eq. (12-8b) to obtain the value of voltage gain and compare the answer with the answer for gain on page 200.
Solution: From Example 12-2, $g_{fs} = 2950 \ \mu\mho$; then $A_v = (2950 \times 10^{-6})$ $(1545) = 4.5$. From page 200, $A_v = 4$.

12-5.4 Input Resistance. The ac input resistance of a common source amplifier can be found by inspection from Fig. 12-10(b):

$$R_{in} = R_G \qquad (12\text{-}9)$$

Again we emphasize the reason for a large value of R_G is to maintain a high input resistance. From Fig. 12-10(b), $R_{in} = R_G = 3.3 \ \text{M}\Omega$.

12-6 Common-Drain Amplifier

If R_L in Fig. 12-8(a) is shorted and V_o is measured across R_s, we have a common-drain amplifier, as shown in Fig. 12-11(a). Now that R_s is the load resistor, a value greater than a few hundred ohms is desirable. However, for large values of R_s, V_{GS} would be large and drain current would be low. To overcome this problem, resistors R_1 and R_2 provide a forward bias voltage to offset some of the reverse bias due to I_DR_s. By having R_2 adjustable, the operating point may be set midway on the load line. R_G preserves high input resistance of the JFET, which would be lowered by R_1 and R_2.

12-6.1 AC Model. Figure 12-11(b) is an ac model of Fig. 12-11(a). As in previous ac models, coupling capacitors and power-supply resistance is negligible at 1 kHz, the frequency of E_i.

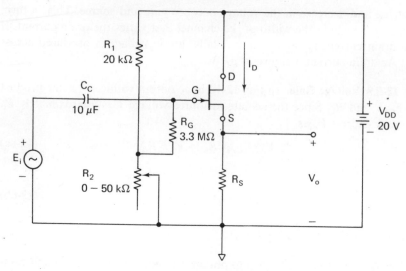

(a) Common-drain amplifier

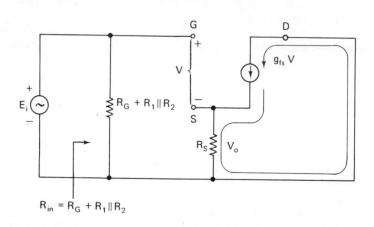

$R_{in} = R_G + R_1 \| R_2$

(b) ac model of circuit (a)

Figure 12-11 Circuit diagram and ac model for a common-drain amplifier.

12-6.2 Voltage Gain. From Fig. 12-11(b) we note that V does not equal E_i. Writing an input loop equation,

$$E_i = V + V_o \qquad (12\text{-}10a)$$

where

$$V_o = g_{fs} V R_s \qquad (12\text{-}10b)$$

Thus

$$E_i = V + g_{fs}VR_s = V(1 + g_{fs}R_s) \qquad (12\text{-}10\text{c})$$

and
$$V = \frac{E_i}{1 + g_{fs}R_s} \qquad (12\text{-}10\text{d})$$

To find voltage gain $A_v = V_o/E_i$, substitution of Eq. (12-10d) into Eq. (12-10b) yields

$$A_v = \frac{V_o}{E_i} = \frac{g_{fs}R_s}{1 + g_{fs}R_s} \approx 1 \qquad \text{if } g_{fs}R_s \gg 1 \qquad (12\text{-}11)$$

Example 12-4: R_2 in Fig. 12-11 is adjusted to set the same operating point as in Fig. 12-7(b), $V_{GS} = -2$ V, $I_D = 4.4$ mA, and $V_{DS} = 11$ V. Calculate (a) source resistor R_s and (b) voltage gain.
Solution:

(a) Writing the output loop equation for Fig. 12-11(a):

$$V_{DD} = V_{DS} + I_D R_s$$

Solving for R_s,

$$R_s = \frac{V_{DD} - V_{DS}}{I_D} = \frac{20 \text{ V} - 11 \text{ V}}{4.4 \text{ mA}} \approx 2.05 \text{ k}\Omega$$

(b) From Example 12-2, $g_{fs} = 2950 \ \mu\text{V}$. Using Eq. (12-11),

$$A_v = \frac{(2950 \times 10^{-6})(2.05 \times 10^3)}{1 + (2950 \times 10^{-6})(2.05 \times 10^3)} = \frac{6.04}{1 + 6.04} = 0.86$$

12-6.3 Input Resistance. The input resistance for a common-drain amplifier is similar to that of a common-source amplifier. That is, input resistance is primarily controlled by R_G. From Fig. 12-11(b),

$$R_{in} = R_G + R_1 \| R_2 \qquad (12\text{-}12\text{a})$$

Since $R_G > R_1 \| R_2$,

$$R_{in} \approx R_G \qquad (12\text{-}12\text{b})$$

For Fig. 12-11(a), $R_{in} \approx 3.3$ MΩ.

12-7 JFET Applications

12-7.1 Chopper Circuit. Amplifying dc signals is usually more of a problem than amplifying ac signals. The reason is that one amplifier must be

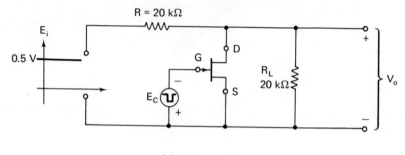

(a) JFET chopper circuit

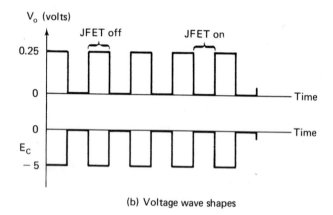

(b) Voltage wave shapes

Figure 12-12 JFET chopper circuit and resultant output wave form.

directly coupled (no capacitor) to the next stage. When stages are directly coupled, a bias voltage change in one stage is amplified by the next stage to change its bias adversely. A chopper circuit converts dc levels to an ac signal, and an amplifying circuit to process ac signals may be used. Figure 12-12(a) is a JFET chopper circuit. Chopper voltage E_c turns the JFET on and off. When the JFET is on, $r_{ds(on)}$ short-circuits R_L and $V_o \cong 0$ V. When the JFET is off, E_i divides between R and R_L and $V_o = E_i/2$. This circuit operates only in the ohmic region; therefore, $r_{ds(on)}$ should be a low value. Figure 12-12(b) shows the resultant wave form.

12-7.2 Sample-and-Hold Circuit. To have a digital computer record, analyze, and store information on an ac voltage, an analog-to-digital conversion system must be used. One part of this system is a sample-and-hold circuit similar to Fig. 12-13. This circuit samples the input signal at definite intervals and holds the input voltage sample long enough for a conversion circuit to convert the signal to a digital signal. In Fig. 12-13 when $V_{GS} = 0$,

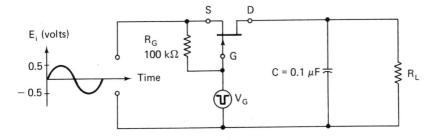

(a) Sample and hold circuit

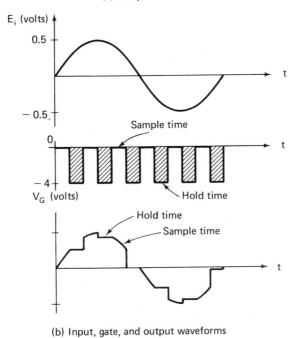

(b) Input, gate, and output waveforms

Figure 12-13 Sample-and-hold circuit and wave forms.

voltage E_i is connected through $r_{ds(on)}$ of the JFET to *hold* capacitor C and C changes to a voltage equal to E_i. Voltage E_c then reverse-biases the gate-to-source junction and disconnects input voltage E_i. C holds its sample voltage during this interval. Figure 12-13(b) shows the wave shapes of input, output, and gate-control voltage.

12-7.3 FET–BJT Amplifier. Figure 12-14 shows an application of both types of transistors, an FET and a BJT. The *pnp* BJT provides current gain for the FET. R_2 is adjusted to set the dc voltage from collector to ground at

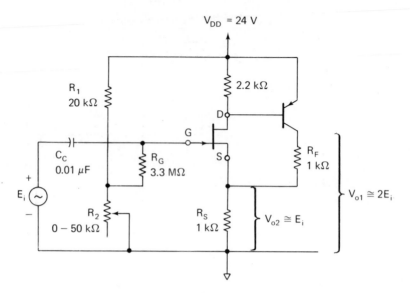

Figure 12-14 FET–BJT amplifier.

$\frac{1}{2}V_{DD}$. Most of the ac drain current enters the base of the BJT. The voltage gain is

$$A_{v1} = \frac{V_{o1}}{E_i} = \frac{R_F + R_s}{R_s} \tag{12-13}$$

If R_F is short-circuited, the expression for voltage gain is

$$A_{v2} = \frac{V_{o2}}{E_i} \approx \frac{R_s}{R_s} = 1 \tag{12-14}$$

Example 12-5: If $R_F = R_s = 1\ k\Omega$ in Fig. 12-14, calculate voltage gain V_{o1}/E_i.
Solution: From Eq. (12-13),

$$A_{v1} = \frac{V_{o1}}{E_i} = \frac{1\ k\Omega + 1\ k\Omega}{1\ k\Omega} = 2$$

12-8 Mode of Operation

The junction field-effect transistor is classified as type A; a subclassification would be *n*-channel or *p*-channel. The category of type A is for a field-effect transistor that operates in the *depletion region*. Figure 12-15 shows the depletion region for a *n*-channel JFET. Chapter 13 introduces the metal-oxide semiconductor field-effect transistors, MOSFET. There are two basic types

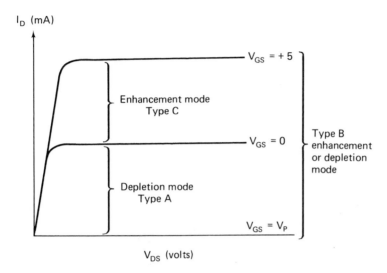

Figure 12-15 Regions of operation for different types of FETs.

of MOSFET, type B and type C, depending on whether or not a conducting channel is built between drain and source. Figure 12-15 also shows the region of operation for both type B and type C FETs. The mode of operation for type C FETs is called the *enhancement mode*.

Problems

12-1 (a) How many charge carriers does a bipolar junction transistor depend on?
(b) How many charge carriers does a field-effect transistor depend on?

12-2 What are the two types of field-effect transistors?

12-3 What type of material is the gate made of in an *n*-channel JFET?

12-4 Name the three connections to a JFET.

12-5 What is the direction of conventional current between drain and source for (a) an *n*-channel JFET; (b) a *p*-channel JFET?

12-6 What controls the width of the channel?

12-7 The current–voltage characteristics for a JFET are a plot of what?

12-8 Give three names for the flat portion of the I_D–V_{DS} characteristics.

12-9 What does the symbol I_{DSS} represent?

12-10 If from a set of I_D–V_{DS} characteristics (similar to Fig. 12-3) $\Delta V_{DS} = 1$ V and $\Delta I_D = 10$ mA, calculate $r_{ds(on)}$.

12-11 What is the voltage that causes the space-charge regions to touch and close called?

12-12 (a) What is the purpose of using R_G in Fig. 12-6(a)?
(b) Why should its value be between 1 and 10 MΩ?

12-13 Refer to Fig. 12-6(b) to operate at $I_D = 5$ mA. What value of R_s is needed to properly bias the JFET?

12-14 In Fig. 12-7(a), let $R_L = 1167\ \Omega$, $R_s = 833\ \Omega$, and $V_{DD} = 16$ V. (a) Plot the load line on Fig. 12-6(b); (b) locate the operating point.

12-15 If a capacitor is put across R_s in Problem 12-14, draw the ac load line. Consider $X_c \approx 0$.

12-16 If C_s is removed in Fig. 12-8(a), determine the peak-to-peak output voltage developed from drain to source.

12-17 Calculate the voltage gain for Problem 12-16.

12-18 Use the characteristic curves of Fig. 12-6(b) to calculate g_{fs} at the operating point of $V_{DS} = 10$ V and $V_{GS} = -2.5$ V.

12-19 For a particular JFET amplifier, $R_L = 1167\ \Omega$ and $g_{fs} = 2500\ \mu\mho$, determine the voltage gain. R_s is shunted by a capacitor.

12-20 If $R_G = 1$ MΩ in Fig. 12-10(a), what is the input resistance of the amplifier?

12-21 Determine the voltage gain for the common-drain amplifier of Fig. 12-11(a) if $R_s = 2$ kΩ and $g_{fs} = 2500\ \mu\mho$.

12-22 Refer to Fig. 12-14; if $R_F = 3$ kΩ and $R_s = 1$ kΩ, calculate voltage gain.

13

MOSFETs and CMOS

13-0 Introduction

In Chapter 12 the junction field-effect transistor (JFET) was shown to operate on the principle that an electric field, established by a *pn* junction, controlled the resistance of a conducting channel between source and drain terminals. It is also possible to establish an electric field that controls channel resistance by substituting a capacitor for the *pn* junction. The resulting device is called a *metal-oxide-semiconductor field-effect transistor*. Common abbreviations are *MOSFET*, *MOS*, and *MOST*. A less familiar but more descriptive name for the MOS transistor is the *insulated-gate field-effect-transistor*, or IGFET. This name was derived from the structure of the device, as will be demonstrated in Section 13-1.

Since complementary MOS integrated circuits have become so important in many fields, this chapter will be concerned primarily with enhancement-mode MOSFETs. Applications will be selected to show how they are employed both in computer or digital circuits and amplifier or linear circuits.

13-1 MOSFET Construction and Circuit Symbols

13-1.1 Construction. MOSFETs are classified first according to whether or not a conducting channel is fabricated between the source and drain terminals. In Fig. 13-1(a) and (b), the *p*-type substrate is a foundation for manufacturing either type of MOSFET. Heavily doped *n*-type material (N^+) is

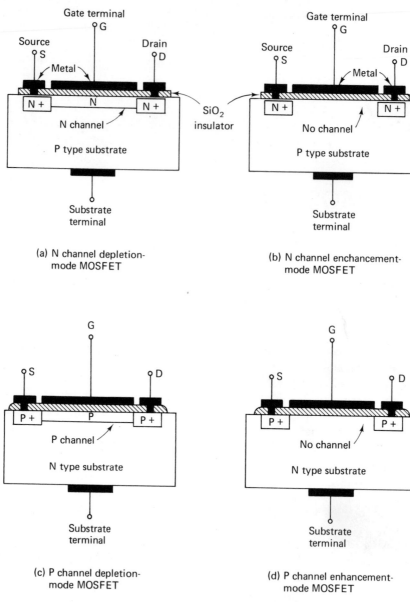

Figure 13-1 Classification of MOSFETs.

diffused onto the substrate to form source and drain areas. These areas are connected via metal ohmic contact pads and bonding wires to source and drain terminals. An *N* channel is diffused between source and drain onto the substrate of a *depletion-mode* MOSFET. No channel is diffused between source and drain of an *enhancement-mode* MOSFET. An insulating layer of silicon dioxide (SiO_2) is next deposited over the surface between source and drain. Finally, a layer of metal is deposited over the insulating layer, and the metal layer is wired to a gate terminal. The metallic gate layer serves as one plate of a capacitor. An insulating layer of silicon dioxide is the capacitor's dielectric. The other capacitor plate is either the N-channel semiconductor material of Fig. 13-1(a) or the P-type substrate of Fig. 13-1(b). Compare the MOSFET and JFET to see that a gate capacitor and gate pn junction, respectively, constitute their most significant difference. Since the leakage current of a capacitor can be very much smaller than the leakage current around a pn junction, we see why the gate input resistance of a MOSFET can be much higher than the normally high gate resistance of a JFET. As a matter of fact, MOSFET gate resistance can reach the spectacular value of 1 million billion ohms!

MOSFETs are also made with N-type substrates and P^+ source and drain. A p-type channel is constructed for depletion-mode MOSFETs and *no* channel is constructed for enhancement-mode MOSFETs. As shown in Fig. 13-1(c) and (d), these two classes of MOSFETs are grouped together by the subclassification of P channel. This subclassification differentiates them from the N-channel MOSFETs in Fig. 13-1(a) and (b). Before studying the differences between depletion and enhancement modes of operation, it is wise to review their classifications.

13-1.2 MOSFET Classification Summary and Symbols.

Electrical symbols for depletion- and enhancement-mode MOSFETs are summarized in Fig. 13-2. The substrate's arrow points toward the N-type semiconductor material,

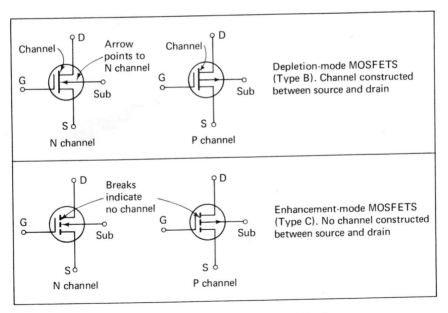

Figure 13-2 MOSFET symbols and classifications.

toward the subterminal for N-type substrate or p-channel, and toward the channel for N-channel. Thus arrowhead direction tells whether the MOSFET is N-channel or P-channel in nature. The vertical channel line symbol is unbroken if the MOSFET is a depletion mode, sometimes referred to as type B. A broken line in the channel symbol indicates that no channel is present and defines an enhancement-mode or type C MOSFET.

13-2 Enhancement-Mode MOSFET Electrical Characteristics

13-2.1 Inducing a Conducting Channel. An N-channel enhancement-mode MOSFET is connected according to the test circuit in Fig. 13-3(a). Voltage V_{GS} between gate and source is zero and only a tiny leakage current of less than 1 nA is conducted between drain and source. Observe that the drain must always be positive with respect to the substrate; if it is not, the drain pn junction can become forward-biased into destruction.

By connecting gate to drain in Fig. 13-3(b), a positive voltage is put on the gate with respect to substrate and source. The gate metal can therefore be shown as accumulating $+$ charges like one plate of a capacitor. On the other side of the SiO_2 insulator, the other capacitor plate must attract minority electrons from the P-type substrate. These electrons line up to form a conducting N channel between source and drain so that drain current can flow. This action can be summarized by the following statement:

> For an N-channel enhancement-mode MOSFET, make the gate voltage *positive* with respect to the source to *induce* an N channel and *enhance* the conduction of drain current I_D.

A corresponding action takes place in a p-channel enhancement-mode MOSFET. Use Fig. 13-3 as a guide; reverse the voltage polarities shown for V_{DS} and V_{GS}; change the substrate to N type, change both source and drain to P^+-type semiconductor material. With zero gate voltage in Fig. 13-3(a), the MOSFET will be nonconducting between drain and source, or cut off. By making the gate negative with respect to the source, holes will be attracted to induce a p-channel between P^+ source and P^+ drain. Current can now flow from source to drain and the MOSFET is described as *on*, because its ability to conduct current has been enhanced.

Figure 13-3(b) shows an important electrical characteristic supplied by manufacturers. They specify a value of drain current when $V_{GS} = V_{DS}$. This value of drain current I_D is symbolized by $I_{D(on)}$. It is shown on the drain characteristics of Fig. 13-5. Figure 13-3(a) shows another electrical characteristic given by manufacturers, I_{DSS}. As in the JFET, I_{DSS} indicates the drain-

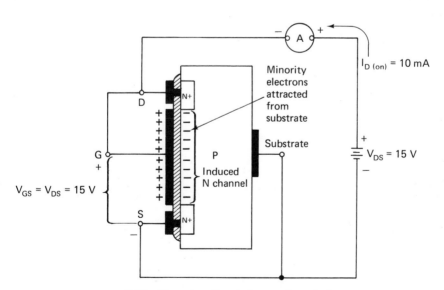

(a) Drain current is essentially zero
with zero gate voltage

(b) Positive gate voltage induces a channel of
electrons to allow drain current flow

Figure 13-3 Inducing a channel of minority carriers enhances flow of
drain current in the *N*-channel enhancement-mode MOSFET.

to-source current that flows when the gate is shorted to the source. It is typically 1 nA for an enhancement-mode MOSFET.

13-2.2 Gate Threshold Voltage. We learned from Section 13-2.1 that with zero gate voltage, I_{DSS} was essentially zero. Also when gate-source voltage V_{GS} equals drain-source voltage V_{DS}, $I_{D(on)}$ is in the milliampere range. Intuitively we reason that there must be some intermediate value of gate-source voltage that will just barely induce a channel. In the test circuit of Fig. 13-4, V_{GS} may be adjusted from $V_{GS} = 0$ V to $V_{GS} = V_{DD} = 15$ V. To measure gate threshold voltage, $V_{GS(th)}$, first set $V_{DD} = V_{DS}$ to 15 V. Initially set $V_{GS} = 0$. Then increase V_{GS} until I_D measures 10 μA. The value of V_{GS} required to set $I_D = 10$ μA is recorded as *gate threshold voltage*, $V_{GS(th)}$. This is the minimum gate voltage needed to barely induce a channel. To measure $V_{GS(th)}$ for a *p*-channel MOSFET, reverse connections to V_{DD}, the ammeter and both voltmeters.

13-2.3 I_D–V_{DS} Drain Characteristic Curves. The test circuit of Fig. 13-4 can also be used to obtain drain characteristic curves for the MOSFET. (These are analogous to collector characteristics of a BJT.) First adjust V_{DS} to zero volts. Then set V_{GS} to a convenient value greater than $V_{GS(th)}$; for example, set $V_{GS} = 8$ V. Next, increase V_{DS} or V_{DD} from 0 V to 15 V in convenient increments and record the corresponding values of drain current, I_D. Then plot each pair of corresponding values for I_D and V_{DS} as a point on

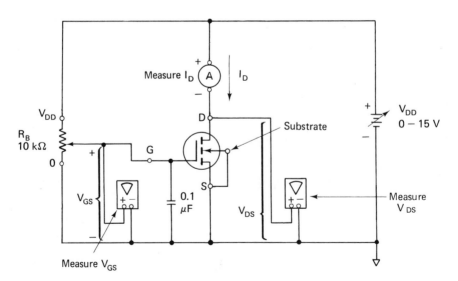

Figure 13-4 Test circuit to measure drain characteristics of an *N*-channel type C MOSFET.

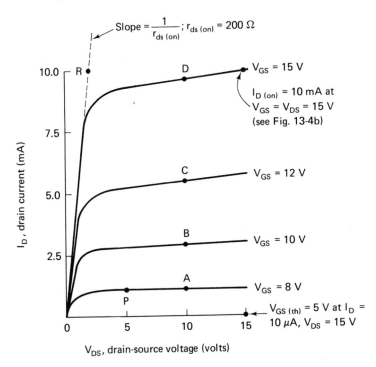

Figure 13-5 I_D–V_{DS} characteristic curves for an N-channel enhancement-mode MOSFET.

a graph, as in Fig. 13-5. For example, point P is located at $I_D = 1$ mA, $V_{DS} = 5$ V. Connecting all points with a smooth line generates the I_D–V_{DS} characteristic curve labeled $V_{GS} = 8$ V in Fig. 13-5. Now change V_{GS} to 10 V and let V_{GS} remain constant while varying V_{DS} between 0 and 15 V. The resulting drain characteristic is labeled $V_{GS} = 10$ V in Fig. 13-5. The procedure is repeated until a family of I_D–V_{DS} drain characteristic curves is obtained.

Observe that for values of V_{DS} greater than a few volts, the drain current stays fairly constant, as long as gate voltage remains constant. The mechanism that limits drain current is the same pinchoff action that occurred in the JFET channel.

13-2.4 Drain-Source "On" Resistance. At values of V_{DS} below roughly 1 V, the I_D–V_{DS} characteristic curves approximate a straight line rising from the origin. Since the I–V characteristic of a resistor is a straight line, we conclude that the MOSFET behaves like a resistor. Thus the rising straight-line portion of each I_D–V_{DS} curve is called the *ohmic region*. Manufacturers give a value for the ohmic resistance represented by the MOSFET at a particular value of gate voltage and at very low drain current (actually $I_D = 0$). This value of ohmic resistance is called *drain source "on" resistance*, $r_{ds(on)}$. It can be approximated fairly accurately from the I_D-V_{DS} characteristic in Fig. 13-5. To measure $r_{ds(on)}$ at $V_{GS} = 15$ V, pick any point on its ohmic region. Observe that the ohmic region has been extended for convenience in Fig. 13-5 by a dashed construction line to point R. The coordinates of point R are $I_D = 10$ mA, $V_{DS} = 2.0$ V. Calculate $r_{ds(on)}$ from

$$r_{ds(on)} = \frac{V_{DS}}{I_D} = \frac{2.0 \text{ V}}{10 \text{ mA}} = 200 \ \Omega \qquad \text{at } V_{GS} = 15 \text{ V} \qquad (13\text{-}1)$$

13-2.5 Forward Transconductance. From inspection of Fig. 13-5 we see that input gate-source voltage V_{GS} controls output drain current I_D. To understand how a MOSFET amplifies, we need to know the relationship between a change in input gate voltage and the corresponding change in output current. The term *transfer characteristic* implies a relationship or ratio between an output term and an input term. If the ratio desired is between current and voltage, the units would be *conductance* or *admittance*. Thus the ratio between a *change* in drain current, ΔI_D, and a *change* in gate voltage, ΔV_{GS}, is appropriately called *transconductance*. Symbols used in the literature and date sheets are g_m, g_{fs}, and y_{fs}. We shall use g_{fs}.

Although it is possible to measure g_{fs} from the drain characteristics of Fig. 13-5, it is more informative to plot an I_D-V_{GS} curve or transfer characteristic for constant V_{DS}. For example, read the coordinates of points A, B, C, and D in Fig. 13-5 in terms of I_D and V_{GS}. Note that $V_{DS} = 10$ V for each point, as shown in Table 13-1. Plot these points in Fig. 13-6. As shown, g_{fs} is evaluated from the slope of the I_D-V_{GS} curve by

$$g_{fs} = \frac{\Delta I_D}{\Delta V_{GS}} \qquad (13\text{-}2)$$

Example 13-1: Evaluate g_{fs} at $V_{GS} = 12$ V from Fig. 13-6.

Solution: Enter the horizontal axis at $V_{GS} = 12$ V and proceed up to its intersection with the I_D-V_{GS} curve at point C. Draw a construction line tangent to the curve at point C and extend it for convenience to the V_{GS} axis. Measure $\Delta I_D = 9.6$ mA and the corresponding $\Delta V_{GS} = 7$ V. From Eq. (13-2),

$$g_{fs} = \frac{\Delta I_D}{\Delta V_{GS}} = \frac{9.6 \text{ mA}}{7 \text{ V}} \approx 1400 \ \mu\mho$$

Table 13-1 I_D-V_{GS} CURVES FOR CONSTANT V_{DS}

Point	V_{GS}, V	I_D, mA	V_{DS}, V
A	8	1.0	10
B	10	2.9	10
C	12	5.5	10
D	15	9.6	10

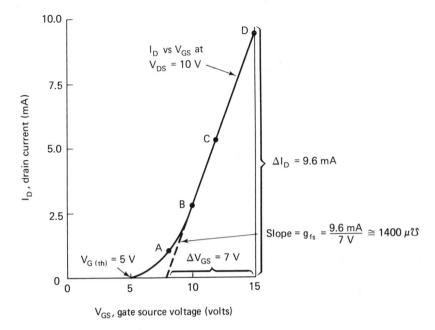

Figure 13-6 I_D–V_{GS} transfer characteristic for the enhancement-mode MOSFET of Fig. 12-5.

13-3 Biasing the Enhancement-Mode MOSFET

13-3.1 Feedback Biasing. There are two basic methods to bias (establish a dc operating point for) enhancement-mode MOSFETs. In the feedback bias arrangement of Fig. 13-7(b), no current flows through resistor R_F. Therefore, the dc voltage levels at drain and gate must be the same with respect to ground. This is because there is zero voltage drop across R_F and we conclude that *any* value of R_F will allow V_{GS} to equal V_{DS}. However, as will be shown in Section 13-4.2, R_F must be large or else ac input resistance will be too low. By locating all points where $V_{GS} = V_{DS}$ on the drain characteristics of Fig. 13-7(a), we plot the locus of $V_{GS} = V_{DS}$. The MOSFET must operate at some point along this feedback bias line. By drawing a load line of $R_L = 3$ kΩ and $V_{DD} = 15$ V for Fig. 13-7(b) onto the drain characteristics, we locate all operating points allowed by the circuit. The only operating point that would satisfy *both* load line and feedback bias line is located by their intersection at point *B*.

Example 13-2: If R_L is changed to 6 kΩ, find V_{DS} and I_D.
Solution: Draw a load line from $V_{DS} = V_{DD} = 15$ V to $I_D = V_{DD}/R_L =$

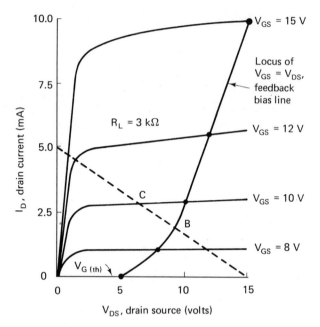

(a) Drain characteristics for N or P channel
enhancement-mode MOSFET

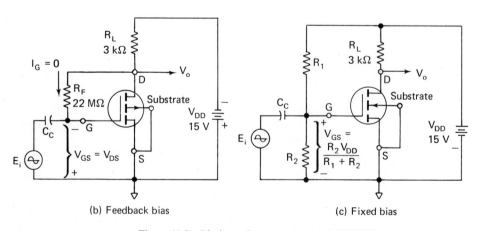

(b) Feedback bias

(c) Fixed bias

Figure 13-7 Biasing enhancement-mode MOSFETs.

15 V/6 kΩ = 2.5 mA. It will intersect the feedback bias line at $V_{DS} \approx 8$ V, $I_D \approx 1.1$ mA.

13-3.2 Fixed Biasing. R_1 and R_2 form a voltage divider for V_{DD} in Fig. 13-7(c) to establish V_{GS} at a fixed value. V_{GS} is expressed by

$$V_{GS} = \frac{R_2}{R_1 + R_2} V_{DD} \qquad (13\text{-}3)$$

In the next example we will investigate how to establish an operating point with fixed bias.

Example 13-3: Choose bias resistors in Fig. 13-7(c) to establish an operating point about halfway up the load line.
Solution: On the load line of Fig. 13-7(a), point C is located about halfway up the load line. Observe that point C lies on the fixed gate-voltage curve of $V_{GS} = 10$ V. Choose a convenient, large value for R_2 of 10 MΩ and solve for R_1 in Eq. (13-3):

$$10 \text{ V} = \frac{10 \text{ M}\Omega}{R_1 + 10 \text{ M}\Omega} 15 \text{ V}, \qquad R_1 = 5 \text{ M}\Omega$$

13-4 Voltage Gain and Input Resistance

13-4.1 Fixed Bias Circuit. Voltage gain V_o/E_i for the MOSFET circuit of Fig. 13-7(c) is identical to its JFET counterpart in Chapter 12. That is,

$$\frac{V_o}{E_i} = g_{fs} R_L \qquad (13\text{-}4)$$

The input resistance R_{in} presented to E_i is simply the parallel combination of bias resistors R_1 and R_2. Resistance between gate and source is so large that the gate is considered an open circuit. Accordingly,

$$R_{in} = R_1 \| R_2 \qquad (13\text{-}5)$$

Example 13-4: In Fig. 13-7(c), g_{fs} is 1400 μ℧ at the operating point. $R_1 = 5$ MΩ and $R_2 = 10$ MΩ. Evaluate (a) voltage gain and (b) input resistance.
Solution:

(a) From Eq. (13-4), $V_o/E_i = (1400 \times 10^{-6} \text{ ℧})(3000 \text{ }\Omega) = 4.2$
(b) From Eq. (13-5), $R_{in} = 5 \text{ M}\Omega \| 10 \text{ M}\Omega = 3.3 \text{ M}\Omega$

From Example 13-4 it is apparent that voltage gains are rather small in MOSFET amplifiers as compared with BJT amplifiers. This is because transconductance of BJT is much higher. Also, input resistance is set by the bias resistors so they should be as large as possible.

13-4.2 Feedback Bias Circuit. Voltage gain for the feedback bias circuit of Fig. 13-7(b) is expressed by

$$A_V = \frac{V_o}{E_i} = g_{fs}(R_F \| R_L) \qquad (13\text{-}6a)$$

But R_F is usually much larger than R_L, so Eq. (13-6a) can be simplified to

$$A_V = \frac{V_o}{E_i} = g_{fs}R_L \qquad (13\text{-}6b)$$

Input resistance is reduced by the Miller effect (see Section 16-6). The gate draws no signal current at medium frequencies. But when E_i goes positive in Fig. 13-7(b) to push current into one side of R_F, V_o goes negative, to pull current from the other side of R_F. Thus R_F draws more current from E_i because of the connection to V_o, and R_F is apparently reduced in size. Because $V_o = A_V E_i$, the resistance presented to E_i is R_{in}, where it can be shown that

$$R_{\text{in}} = \frac{R_F}{1 + A_V} \qquad (13\text{-}7)$$

Example 13-5: In Fig. 13-7(b), $g_{fs} = 1400 \ \mu\text{℧}$ and $R_F = 22 \ \text{M}\Omega$. Find (a) voltage gain and (b) input resistance.
Solution:

(a) From Eq. (13-6b), $A_V = (1400 \times 10^{-6} \ \text{℧})(3000 \ \Omega) = 4.2$

(b) From Eq. (13-7), $R_{\text{in}} = \dfrac{22 \ \text{M}\Omega}{1 + 4.2} = 4.2 \ \text{M}\Omega$

13-5 Depletion-Mode MOSFETs (Type B)

13-5.1 Similarities to the Depletion-Mode JFET. Since the MOSFET in Fig. 13-8 contains an N channel, it must conduct a saturation current I_{DSS} when $V_{GS} = 0$, just like the JFET. Also, like the JFET, the same pinchoff mechanism will limit I_{DSS} to a finite value. If the same polarity of voltage is applied to the gate (with respect to source) as the polarity of channel majority carriers, some majority carriers will be repelled into the substrate. For example, the negative gate voltage in Fig. 13-8(a) repells electrons from the N channel into the substrate. The N channel narrows since it is partially depleted

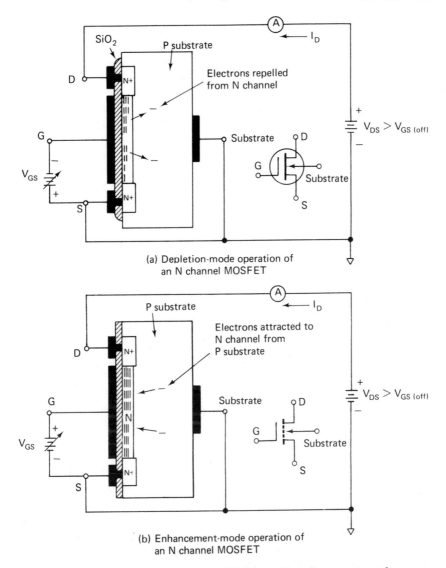

(a) Depletion-mode operation of
an N channel MOSFET

(b) Enhancement-mode operation of
an N channel MOSFET

Figure 13-8 Type B, *N*-channel depletion mode–enhancement mode
MOSFET operation. For *P*-channel MOSFETs, reverse connections to
V_{DS} and V_{GS}.

of carriers. And just as the JFET channel can be pinched off with sufficient
gate voltage, the MOSFET can also be pinched off with sufficient gate voltage.
This gate-source pinchoff or cutoff voltage is represented on data sheets by
the symbol $V_{GS(off)}$. It is equivalent to V_p for the JFET.

13-5.2 Enhancement-Mode Operation of an *N*-Channel MOSFET. It is possible to operate a JFET in the enhancement mode provided that the gate-source voltage is *not* sufficient to forward-bias the gate *pn* junction. For example, if the gate junction is forward-biased by 0.6 V, the gate resistance will drop to a very low value and defeat the purpose of using an FET.

The type B MOSFET may also be operated in the enhancement mode, as in Fig. 13-8(b). When the gate is made positive with respect to the source, electrons are attracted into the *N* channel from the *P* substrate to enhance channel conduction. This action is similar to that of the type C enhancement-mode-only MOSFET. Note that enhancing gate voltages can be much larger for the MOSFET than the JFET because of the gate insulation.

13-5.3 Characteristic Curves for the Depletion Mode–Enhancement Mode MOSFET. By varying V_{DS} from 0 to 20 V in Fig. 13-8(a) and holding V_{GS} = 0, we could record corresponding values of drain current and plot the I_D − V_{DS} curve labeled $V_{GS} = 0$ in Fig. 13-9(a). The sequence can be repeated for $V_{GS} = -1$ V and again for $V_{GS} = -2$ V to plot the corresponding curves in Fig. 13-9(a). Finally, when V_{GS} was increased to -4 V, I_D would be depleted down to 10 μA and we can record gate cutoff voltage $V_{GS(off)}$ as equal to -4 V. This would identify *depletion-mode operation* in the area so designated in Fig. 13-9(a).

By reversing V_{GS} as in the circuit of Fig. 13-8(b), we can take data to plot I_D–V_{DS} characteristics for *enhancement-mode operation*. These curves are identified by $+V_{GS}$ notations in Fig. 13-9(a). By using the technique described in Section 13-2.5, a transfer characteristic can be plotted in Fig. 13-9(b) for $V_{DS} = 10$ V in Fig. 13-9(a). $V_{DS(off)}$, I_{DSS}, and g_{fs} are readily identifiable on this transfer characteristic.

13-5.4 Biasing Techniques for the Depletion Mode–Enhancement Mode MOSFET. Bias techniques for the type B MOSFET are shown in Fig. 13-10. Operating points for Fig. 13-10(a), (b), and (c) are identified by corresponding letters *A*, *B*, and *C* in Fig. 13-9(a). $V_{GS} = 0$ V in Fig. 13-10(a), so I_D is automatically biased at I_{DSS}. In Fig. 13-10(b) self-biasing is used, as with a JFET for depletion operation. Finally, in Fig. 13-10(c) a fixed bias is employed, in the fashion introduced by Fig. 13-7(c).

Type B MOSFETs may be used for the same applications performed by JFETs. The analysis and design procedures are the same as presented in Chapter 12. They also may be biased for enhancement operation, like type C MOSFETs, but the advantage of zero drain current for zero gate voltage is lost. Analysis and design procedures are the same as in Sections 13-3 and 13-4 and will not be repeated. Instead, attention will be directed to the most important and growing areas now being served by MOSFETs.

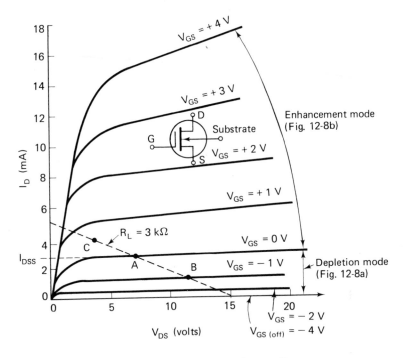

(a) Drain I_D-V_{DS} characteristics for type B
depletion-enhancement mode MOSFET

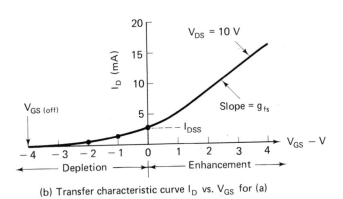

(b) Transfer characteristic curve I_D vs. V_{GS} for (a)

Figure 13-9 Characteristic curves for an *N*-channel depletion mode–
enhancement mode MOSFET (type B).

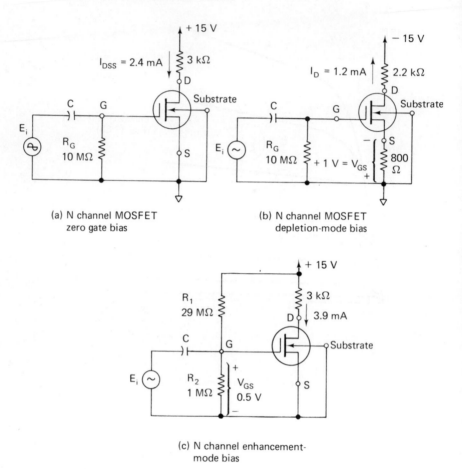

(a) N channel MOSFET
zero gate bias

(b) N channel MOSFET
depletion-mode bias

(c) N channel enhancement-
mode bias

Figure 13-10 Bias circuits for depletion mode–enhancement mode MOSFETs (type B). Operating points for (a), (b), and (c) are shown in Fig. 13-9 as A, B, and C, respectively.

13-6 Introduction to CMOS

Complementary metal-oxide semiconductor (CMOS), or complementary symmetry MOS (COSMOS), identifies arrays of N-channel and P-channel enhancement-mode field-effect transistors located on a single substrate. Although CMOS technology seems to have burst upon the scene only recently, it was actually pioneered by RCA in the 1960s with their COSMOS CD4000 family of digital logic, integrated circuits. At present CMOS circuits are used in electronic watches, pacemakers, and automobile clocks. CMOS technology has made available today a host of low-cost hand calculators which have a calculating ability which exceeds that of massive computers made less than three decades ago. In the fields of automotive, medical, and industrial control, CMOS applications are expanding. It promises to supplant many existing junction transistor technologies and also to make possible the economical solution of problems that could not be solved by existing technologies.

The most significant advantages of CMOS are: (1) A CMOS integrated circuit can be fabricated with about one-fifth of the steps required for a junction transistor circuit; and (2) under certain conditions a CMOS integrated circuit uses less than $\frac{1}{1000}$ of the power used by any other type of integrated circuit. Remember that you pay for your power four times: (1) at the time it is generated, when the waste portion pollutes the environment; (2) when you use it; (3) when you get rid of it; and (4) when the residue again pollutes the environment. To learn why CMOS has such inherent advantages we examine the basic circuit element, which consists of a totem pole or cascade arrangement with two complementary MOSFETs.

13-7 Basic CMOS Circuit Element— The Inverter

13-7.1 Circuit and MOSFET Characteristics. In Fig. 13-11(c) the upper *P*-channel MOSFET is connected as a load (instead of load resistor R_L) to the lower *N*-channel MOSFET. When two transistors are connected in series, as in this figure, it is called a *complementary pair*. For brevity call the upper MOSFET PMOS and the lower MOSFET NMOS. The (+) supply terminal is labeled V_{DD} and the ground-supply terminal is labeled V_{SS}, according to existing convention. Since the drain terminals are in series, the same drain current must flow in both transistors.

A family of drain characteristics for PMOS is shown in Fig. 13-11(a). Observe how V_{DSP} and V_{GSP} have negative values, because both drain and gate are negative with respect to reference source terminal S_P. In Fig. 13-11(b) drain characteristics for NMOS are shown with positive V_{DSN} and V_{GSN} because both drain and gate will be positive with respect to reference source terminal S_N.

There is a definite purpose for reversing the normal orientation of I_D–V_{GS} curves in Fig. 13-11(a) so that they rise from right to left. Observe that if we had a resistor load on NMOS we would draw the load line (*I–V* resistor characteristic) on Fig. 13-11(b) to rise from right to left, with a slope of $1/R_L$. Since PMOS is a load on NMOS, the stage is being set to draw the PMOS *I–V* characteristic as a load on NMOS in Section 13-7.2.

13-7.2 Inverting a Low Input Voltage. In Fig. 13-12(a), input voltage E_i = 0 V. NMOS has zero gate voltage. PMOS has an enhancing gate voltage

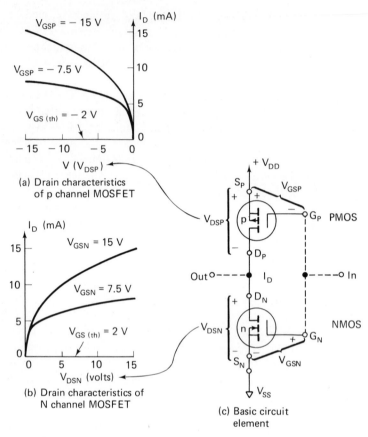

Figure 13-11 Drain characteristics for the basic complementary CMOS-circuit MOSFETS in (c) are shown in (a) and (b).

of 15 V. Therefore, PMOS conducts and NMOS is turned off. No drain current can flow through NMOS. If another CMOS gate was connected as a load on the output terminal, the high gate impedance would draw zero drain current from PMOS. However, PMOS could furnish several milliamperes to any resistor load between output and ground. A simplified model of the CMOS pair is shown in Fig. 13-12(c) with an added resistor load to show how PMOS is modeled by a 200-Ω to 500-Ω resistor, and NMOS is modeled by an open circuit. Since PMOS can supply load current, the CMOS complementary pair is said to have an ability to *source* current.

A more precise determination of operating-point location is obtained from Fig. 13-12(b). NMOS must operate along its gate bias curve $V_{GSN} = 0$ V. PMOS must operate along its gate bias curve of $V_{GSP} = -15$ V. Since

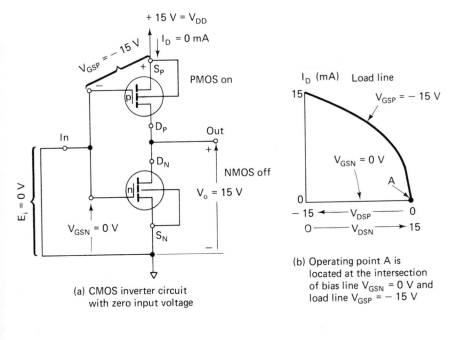

(a) CMOS inverter circuit with zero input voltage

(b) Operating point A is located at the intersection of bias line $V_{GSN} = 0$ V and load line $V_{GSP} = -15$ V

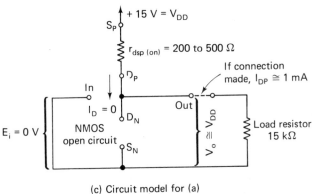

(c) Circuit model for (a)

Figure 13-12 The CMOS pair in (a) is modeled in (c) and has the operating-point location shown in (b).

PMOS is the load on NMOS, we draw its characteristic curve in reverse (just as we did R_L in Section 5-7). The only operating point that will satisfy both characteristic curves ($V_{GSN} = 0$ V, $V_{GSP} = -15$ V) is their intersection at operating point A. From point A we read $I_D = 0$, $V_{DSP} = 0$, and $V_{DSN} = 15$ V. Since V_{DSN} is also output voltage V_o, $V_o = 15$ V. If no load resistor is connected to the output terminal, V_o will be only a few millivolts less than

V_{DD}. Since output voltage V_o is high ($+15$ V) when input voltage E_i is low (0 V), the circuit is called an *inverter* because it inverts the input signal.

13-7.3 Inverting a High Input Voltage. Voltage E_i is applied to the inverter input of Fig. 13-13(a) with a value equal to supply voltage V_{DD}. NMOS has an enhancing gate voltage of $+15$ V. Thus NMOS has an enhanced channel and may be modeled by a resistor of 200 to 500 Ω in Fig. 13-13(c). Both gate and source of PMOS are at $+15$ V with respect to ground, and therefore $V_{GSP} = 0$. No P channel has been created, so drain current I_D must be zero.

It is possible to connect a load resistor in series with output terminal and an external supply voltage as in Fig. 13-13. NMOS can conduct a drain current under this condition, so the circuit is described as having the ability to *sink* (conduct) current.

Operating point C is located precisely in Fig. 13-13(b) at the intersection of bias line $V_{GSN} = 15$ V and load line $V_{GSP} = 0$ V. Since $V_{DSN} = 0$ V is also equal to V_o, we again see inverter action where a high input signal ($E_i = 15$ V) is inverted to a low output signal $V_o = 0$ V. After comparing Fig. 13-12(b) with Fig. 13-13(b), it is natural to wonder what path is traveled by the operating point as it moves from point A to point C. Since point A is set by $E_i = 0$ V and C by $E_i = 15$ V, we choose an intermediate voltage of $E_i = 7.5$ V to find the operating path in the next section.

13-7.4 Inverter Operating Path. Assume that E_i is increased from 0 V in Fig. 16-12 to the value of 7.5 V shown in Fig. 13-14(a). V_{GSN} will sweep *up* from $V_{GSN} = 0$ V to $V_{GSN} = 7.5$ V. Simultaneously, V_{GSP} will sweep *down* from -15 V to -7.5 V, as shown in Fig. 13-14. The operating point moves from point A in Fig. 13-12(b) to point B in Fig. 13-14(b). Movement of the operating point is seen by visualizing the intersections of the bias and load lines as they cross swiftly like two bent swords. One sword sweeps down (V_{GSP}) as the other (V_{GSN}) sweeps up. The actual movement of the operating point is shown in Fig. 13-14(c).

As E_i moves positive from its lower limit of 0 V to its upper limit of $+15$ V, the operating point moves over the path ABC in Fig. 13-14(c). Typical threshold voltages are about 2 V for both PMOS and NMOS. So when E_i is greater than 2 V and less than $15 - 2 = 13$ V, both PMOS and CMOS will be partially conducting. Maximum conduction occurs when $E_i \approx V_{DD}/2$, where $I_{DN} = I_{DP} = 7$ mA. This value of current is called *throughput current*. For digital logic circuits it represents a waste of power. Digital circuits will be discussed in Section 13-8. But this apparent disadvantage becomes an advantage when the CMOS inverter is employed as a linear amplifier in Section 13-9.

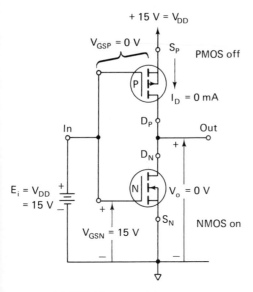

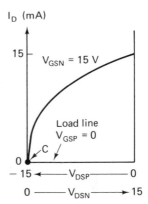

(a) CMOS inverter with
maximum positive input voltage

(b) Operating point C is located
at the intersection of bias
line $V_{GSN} = 15$ V and load
line $V_{GSP} = 0$ V

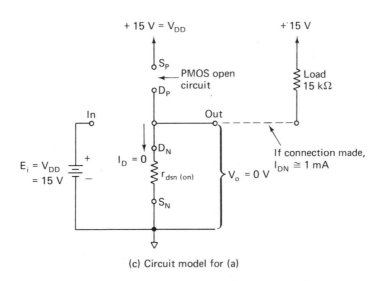

(c) Circuit model for (a)

Figure 13-13 CMOS inverter operation with positive input voltage.

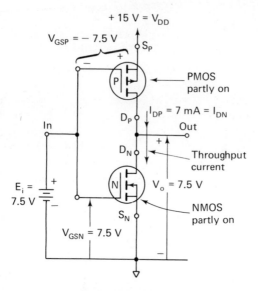

(a) CMOS inverter with
midpoint input voltage

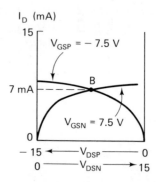

(b) Operating point B is located
at the intersection of bias
line $V_{GSN} = 7.5$ V and load
line $V_{GSP} = -7.5$ V

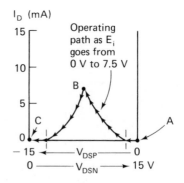

(c) Operating-point path for $E_i = 0$ to 7.5 V
is shown by A to B, and for $E_i = 7.5$ to 15 V
by B to C

Figure 13-14. Operating path of a CMOS inverter.

13-8 CMOS Digital Logic Circuits

13-8.1 Introduction to Logic. In computer or industrial control circuits it is mandatory that signal voltage levels or devices operate in only two possible operating modes. For example, it is easy to tell if a light bulb is on or off. The light output might vary with lamp age and voltage level, but an on lamp is still very different from an off lamp. In computer and control circuits that employ CMOS, the output or input voltage levels are maintained at either $+15$ V or 0 V, corresponding to a high (H) voltage level V_{DD} and a low (L) voltage level (V_{SS} = ground potential), respectively. Rather than use the terms "on" and "off" to describe voltage levels, it is standard practice to use binary system notation 1 and 0. This allows CMOS circuits to be analyzed by the terminology of logic used in computer and control literature. CMOS circuits are usually analyzed with positive logic. That is, the more positive or High-voltage level ($+15$ V) represents 1 and the less positive or Low-voltage level (0 V) represents 0. These ideas are summarized in Table 13-1.

Table 13-1 POSITIVE LOGIC AND VOLTAGE LEVELS

Voltage Level	Electrical Level	Positive Logic Representation
$+15$	H	1
0	L	0

13-8.2 Logical Analysis of a CMOS Inverter. Square-wave input voltage, E_i, in Fig. 13-15(a) applies a logic 1 signal to the inverter from time 0 to 1 ms and a logic 0 signal from time 1 ms to 2 ms. Output V_o develops a logic 1 signal for logic 0 inputs and logic 0 for logic 1 inputs. Thus, as its name implies, an *inverter* inverts a logic signal. In logic language inverting is also called *complementing*.

If E_i is varied slowly in Fig. 13-15(a) from 0 to 15 V and its values are plotted with corresponding values of V_o, a *voltage transfer curve* will result, as shown in Fig. 13-15(b). Here we see that any value of E_i between 0 and 4.5 V (E_i = logic 0) will hold V_o close to its logic 1 value of $+15$ V. Any value of E_i between 10.5 V and 15 V (E_i = logic 1) will hold V_o close to its logic 0 value of 0 V. The remaining range of E_i, from 4.5 to 10.5 V, should be avoided for logic operation because the relation between E_i and V_o is no longer logical, but linear. Operation in the linear region will be explored in Section 13-9. Finally, a standard symbol for the inverter is shown in Fig. 13-15(c). The arrow shape shows the direction of signal flow and symbolizes amplification. The open circle symbolizes inversion.

13-8.3 CMOS NOR Gate. *P*-channel MOSFETs are connected in series and *N*-channel MOSFETs are connected in parallel in the schematic for

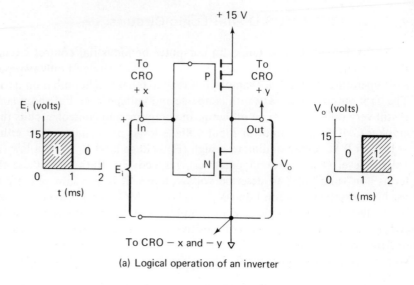

(a) Logical operation of an inverter

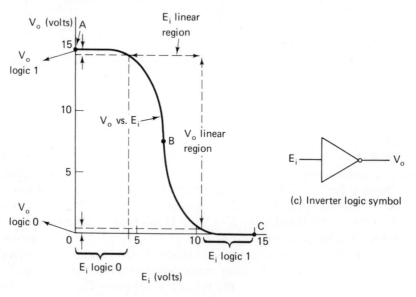

(b) Voltage-transfer characteristic of an
inverter; points A, B, and C correspond
to Figs. 13-12, 13-13, and 13-14, respectively

(c) Inverter logic symbol

Figure 13-15 Analysis of a CMOS inverter.

an integrated-circuit (IC) digital CMOS circuit of Fig. 13-16(a). There are two input terminals, *A* and *B*, but only one output terminal. If a logical 1 signal is applied to input *A* ($E_{iA} = 15$ V), N_A is enhanced *on* and V_o is at logic 0 ($V_o = 0$ V). If input *B* is at logical 1, N_B is enhanced *on* and clamps V_o to logic 0. Thus if either *A* or *B* are at 1, V_o must be at 0. The only way for V_o

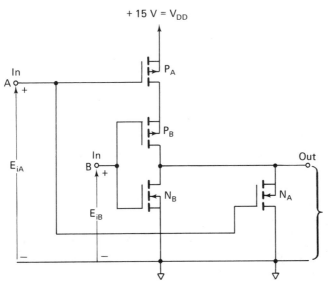

(a) CMOS NOR gate

In		Out
E_{iA}	E_{iB}	V_o
0 V	0 V	15 V
0	15	0
15	0	0
15	15	0

In		Out
A	B	Out
0	0	1
0	1	0
1	0	0
1	1	0

(b) Voltage-level and logic-level truth tables for (a)

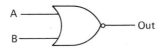

(c) Logic symbol for (a) and (b); NOR gate

Figure 13-16 Analysis of a CMOS NOR gate.

to be at logical 1 is to apply logic 0 to *both* A and B. With zero volts on both inputs, both N_B and N_A are *off*. With $+15$ V on A, P_A is enhanced *on*, connecting $+15$ V from V_{DD} through its low channel resistance to the source terminal of P_B. V_{GS} of P_B is -15 V, so P_B is enhanced *on*. Since both P_A and P_B are *on*, V_{DD} is essentially connected to the output terminal, placing V_o at $+15$ V.

Corresponding input–output conditions described above are summarized in the *truth table* of Fig. 13-16(b). It would be time-consuming to redraw the circuit of Fig. 13-16(a). Since it is a widely used circuit arrangement, it may be represented by the simpler symbol in Fig. 13-16(c). The name of this circuit is derived from the logic truth table. A 1 O͟R a 1 at the inputs does N͟ot give a 1 at the outputs. Underlined letters yield the name NOR. The name *gate* is also applied to this and similar types of logic circuits. It means that the circuit acts like a gate that lets a 1 appear at the output only under particular input conditions. For example, a NOR gate allows a 1 output only when all inputs are 0.

13-8.4 CMOS NAND Gate. Parallel-wired *P*-channel MOSFETs and series-wired *N*-channel MOSFETs yields the NAND gate logic element of Fig. 13-17(a). Logic 0 levels (0 V) at either input A or B enhance channels in P_A or P_B, respectively, to establish output voltage V_o at logic 1. That is, V_{DD} is connected through conducting source-drain channels of either P_A or P_B to V_o.

Channels will be enhanced in N_A and N_B only when both inputs A and B have simultaneous 1 signals. With N_A and N_B enhanced, V_o is grounded through $r_{ds(on)}$ of both N_A and N_B in series. Thus V_o will only be at logic 0 when *both* In A and In B are at logic 1. These logic input–output relations are summarized in the truth tables of Fig. 13-17(b). We see that a 1 A͟ND a 1 at each input does N͟ot give a 1 at the output. The underlined characters show how the term "N͟AN͟D" is derived to describe this particular gate circuit.

13-9 CMOS Linear Operation

In order to operate on the linear region of Fig. 13-15(b), we must first bias the complementary pair at operating point B. A single feedback bias resistor R_F is wired between output and input in Fig. 13-18. R_F can be any value from 100 kΩ to 22 MΩ. Regardless of the value of R_F, the dc output voltage at pin 12 will be $V_{CC}/2$. The voltage-transfer characteristic is drawn on broken scales in Fig. 13-18(b) in order to show a peak-to-peak input

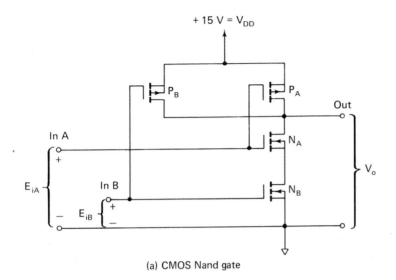

(a) CMOS Nand gate

In		Out
E_{iA}	E_{iB}	V_o
0 V	0 V	15 V
0	15	15
15	0	15
15	15	0

In		Out
A	B	Out
0	0	1
0	1	1
1	0	1
1	1	0

(b) Voltage-level and logic-level truth tables for (a)

(c) Logic symbol for (a) and (b); Nand gate

Figure 13-17 Analysis of a CMOS NAND gate.

voltage swing of $E_i = 20$ mV. The resulting peak-to-peak output voltage swing is 1.0 V, so voltage gain A_V is

$$A_V = \frac{V_o}{E_i} = \frac{1.0 \text{ V}}{0.020 \text{ V}} = 50$$

Input resistance is given by Eq. (13-7), and for $R_F = 22$ MΩ, $R_i = R_F/50 \approx 0.5$ MΩ.

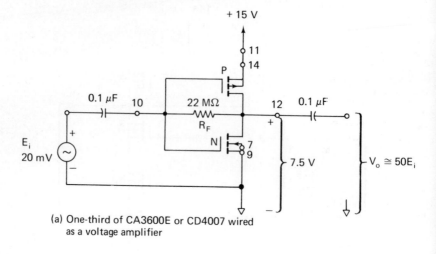

(a) One-third of CA3600E or CD4007 wired
   as a voltage amplifier

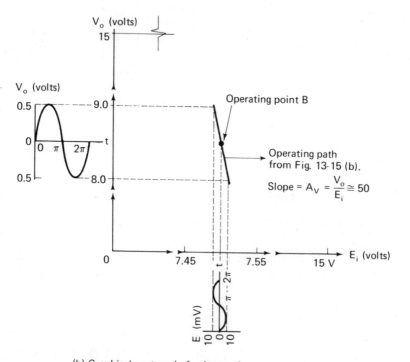

(b) Graphical portrayal of voltage gain

Figure 13-18   Linear operation of a CMOS inverter.

13-10 CMOS Operating Precautions

Voltage levels at input or output terminals should never be allowed to go more positive than the + supply terminal or more negative than the — supply (usually ground) terminal. Power must always be connected to the CMOS circuit *before* control voltages are applied to input or output terminals. Otherwise, permanent damage may result. It follows that devices should never be plugged into a circuit that is powered on.

Since each gate has an extremely high input resistance, an unused gate terminal cannot be left unconnected. It must be connected to one of the power supply busses, V_{DD} or ground, depending on circuit requirements. Otherwise, the gate terminal will accumulate charge on the gate capacitor to bias the CMOS pair *on*, and either waste power or interfere with circuit operation.

Early MOS devices had high failure rates because static charges on the gate capacitor could build up, owing to circuit transients or improper handling, to a value large enough to rupture the gate oxide. Most present-day MOS devices have back-to-back Zeners connected internally between gate and source terminal. When gate voltage builds up beyond the Zener breakdown voltage, the Zener conducts and holds the gate potential at a safe value. These protective Zeners are not usually shown on the schematic diagrams. The Zeners will protect against in-circuit transients of up to a hundred volts or so. However, the human body has a capacitance of roughly 300 pF and, by walking across a rug or waxed floor, can build up a static voltage of between 1 and 15 kV. This amount of energy is too much for even the Zener protection. Therefore, it is sound practice to ground your hands periodically against a chassis (that has a third-wire safety ground). Also all test jigs should be grounded, and handling of CMOS components should be minimized. MOSFET terminals are shorted together by a ring for shipment and IC chips are plugged into a black conducting plastic foam. (The IC chips should never be stored in plastic trays or plastic snow.)

Finally, soldering-iron tips or tools should be grounded to prevent gate damage.

Problems

13-1 What is the single most significant difference in construction between a JFET and a MOSFET?

13-2 What device has the highest input resistance BJT, JFET, or MOSFET?

13-3 On a circuit schematic, how can you tell if the MOSFET is of depletion or enhancement type?

13-4 In which direction does the arrowhead point on the symbol for an *N*-channel MOSFET?

13-5 To induce a conducting channel in an N-type enhancement MOSFET, should the gate be made $(-)$ or $(+)$ with respect to the source?

13-6 If gate and source of a depletion-type MOSFET are shorted together, what will be the value of the drain current?

13-7 If gate and source of an enhancement-type MOSFET are shorted together, will drain current flow?

13-8 What changes must be made in Fig. 13-3 to show operation of a P-channel, enhancement-mode MOSFET?

13-9 An operating point of a MOSFET is located in the ohmic region of Fig. 13-5 at $V_{DS} = 0.1$ V, $I_D = 0.5$ mA. Find $r_{ds(on)}$.

13-10 Find MOSFET transconductance if a gate-voltage change of 2 V causes a drain-current change of 1.9 mA.

13-11 In Fig. 13-7(b), R_F is changed to 10 MΩ. What change results in the operating point?

13-12 If R_L is changed to 2 kΩ and $V_{DD} = 10$ V in Fig. 13-7, find V_{DS} and I_D.

13-13 In Fig. 13-7(c), $R_2 = 8$ MΩ and $R_1 = 7$ MΩ. Find V_{GS} and I_D.

13-14 In a circuit similar to Fig. 13-7(b), $g_{fs} = 2000$ $\mu\mho$, $R_L = 5$ kΩ, and $E_i = 10$ mV. Find V_o.

13-15 What is the input resistance of Fig. 13-7(c) if $R_1 = R_2 = 10$ MΩ?

13-16 What is the input resistance for Problem 13-14 if $R_F = 11$ MΩ?

13-17 If the 800-Ω source resistor in Fig. 13-10(b) is reduced, does I_D increase or decrease?

13-18 In Fig. 13-15(a), $E_i = 15$ V, with the polarity shown. Are the MOSFETs on or off?

13-19 If $E_i = 0$ V (short circuit) in Fig. 13-15(a), are the MOSFETs on or off?

13-20 The output V_o of the NOR gate in Fig. 13-16 is wired to the inverter input of Fig. 13-15. Show the truth table for A and B NOR gate inputs and inverter output.

13-21 Connect the NAND gate's output of Fig. 13-17 to an inverter input. What is the resultant truth table from NAND input to inverter output?

14

Unijunction Transistors

14-0 Introduction

The unijunction transistor (UJT) and programmable unijunction transistor (PUT) are three-terminal semiconductor devices. Both have two stable states of operation, the *on state* and the *off state*. In the off state, current through the device is very small; in the on-state, current through the device is much larger, so the UJT and PUT act as semiconductor switches that either connect or disconnect current to a load. These devices are very special types of switches in that they abruptly trigger from an off state to an on state only when signaled to do so by a control voltage of the proper amplitude. It would be proper to describe them as voltage-sensitive switches.

The amplitude of the control voltage required to switch the UJT and PUT from off to on is called *peak-point voltage*. Peak-point voltage for the UJT is approximately equal to two-thirds of the dc power supply voltage. Peak-point voltage for the PUT can be selected at any convenient value. That is, the circuit designer can program the peak-point voltage for the value best suited for the problem at hand. These desirable characteristics of the UJT and PUT enable them to perform economically in three broad ranges of applications: (1) They can detect when voltage at a particular point reaches a predetermined value, (2) they can measure elapsed time, and (3) they can generate sawtooth waves or pulse signals by performing as the control element in a relaxation oscillator. We begin our study of the UJT and PUT with a look at their unique electrical characteristics.

241

14-1 Principles and Operation of a UJT

14-1.1 Structure. Figure 14-1(a) shows a cross-sectional view of a uni-junction transistor. The UJT is made by diffusing p-type material into bar-shaped n-type material. The diffusion is not in the center of the bar but about 70% from one end. Although a UJT has only one pn junction, its structure is not similar to that of a diode (refer to Fig. 2-5). Also, the UJT is a three-terminal device; a diode is a two-terminal device.

A typical UJT package is shown in Fig. 14-1(b) and is similar to that of a small-signal transistor package. The UJT circuit symbol is given in Fig. 14-1(c). The terminals of a UJT are called emitter, E, base 1, B_1, and base 2, B_2. The ohmic contacts connected to these three terminals perform the same function as any semiconductor ohmic contact discussed in Section 1-8.

The arrowhead of the emitter (1) points toward the n material, (2) indicates the direction of conventional emitter current, and (3) points toward base 1, B_1.

14-1.2 Interbase Resistance. *Interbase resistance*, r_{BB}, is the resistance mea-sured between base 1 and base 2 with the emitter open. This resistance may be measured by an ohmmeter, as shown in Fig. 14-2(a). A resistance bridge may also be used to measure r_{BB}. The internal voltage of the ohmmeter or resistance bridge should be less than 5 V. The ohmmeter reading of Fig. 14-2(a) indicates that the interbase resistance $r_{BB} = 8 \times 1000 \, \Omega = 8 \, k\Omega$. Typical values of r_{BB} range from 4 to 10 $k\Omega$. The resistance from the emitter junction to base 1 is r_{B1} and the resistance from the emitter junction to base 2 is r_{B2}. Their sum is r_{BB},

$$r_{BB} = r_{B1} + r_{B2} \tag{14-1}$$

The UJT may be modeled by two resistors plus a diode, as shown in Fig. 14-2(b).

14-1.3 Intrinsic Standoff Ratio. Intrinsic standoff ratio, η, is the ratio of r_{B1} to r_{BB}. In equation form,

$$\eta = \frac{r_{B1}}{r_{BB}} \tag{14-2}$$

Typical values of η are usually between 0.4 and 0.8. In Fig. 14-3, V_{BB} is applied from B_1 to B_2. By using the voltage divider law the voltage across r_{BB} is

$$\frac{r_{B1}}{r_{B1} + r_{B2}} V_{BB} = \frac{r_{B1}}{r_{bb}} V_{BB} = \eta V_{BB} \tag{14-3}$$

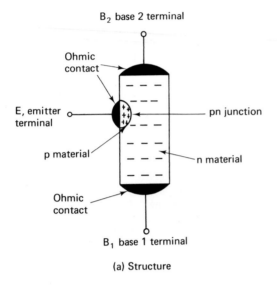

(a) Structure

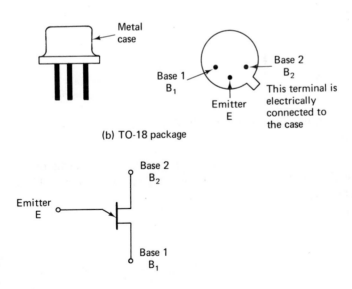

(b) TO-18 package

(c) Circuit symbol

Figure 14-1 Unijunction transistor.

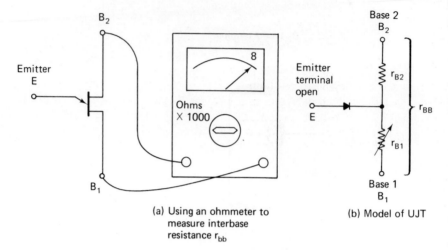

(a) Using an ohmmeter to measure interbase resistance r_{bb}

(b) Model of UJT

Figure 14-2 Measurement and model of a UJT showing interbase resistance r_{BB}.

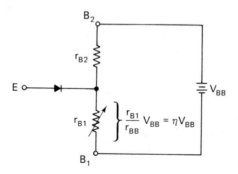

Figure 14-3 Model of a UJT showing intrinsic standoff ratio, η.

Note that in our present discussion no voltage is connected to the emitter terminal.

Example 14-1: The manufacturer of a particular UJT gives the value of η to be 0.6. If interbase resistance r_{BB} is 8 kΩ, calculate (a) r_{B1}, (b) r_{B2}, and (c) the voltage across r_{B1} if $V_{BB} = 10$ V.
Solution:

(a) Rearranging Eq. (14-2),

$$r_{B1} = \eta r_{BB} = (0.6)(8 \text{ k}\Omega) = 4.8 \text{ k}\Omega$$

(b) From Eq. (14-1),

$$r_{B2} = r_{BB} - r_{B1} = 8 \text{ k}\Omega - 4.8 \text{ k}\Omega = 3.2 \text{ k}\Omega$$

(c) Using Eq. (14-3),

$$\text{voltage across } r_{B1} = \eta V_{BB} = (0.6)(10 \text{ V}) = 6 \text{ V}$$

14-1.4 Theory of Operation. The UJT may be operated in one of two states, an off state and an on state. The key to using the UJT is understanding how we can go from one state to the other. Figure 14-4 is needed to study the principles of triggering the UJT from the off state to the on state.

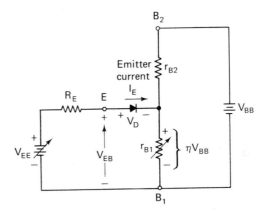

Figure 14-4 Triggering a UJT to its on state.

In Fig. 14-3 the emitter diode is reverse-biased. To overcome this reverse bias, V_{EE} in Fig. 14-4 must be increased to a value large enough to forward-bias the emitter diode by approximately 0.5 V ($V_D \approx 0.5$ V). When V_{EE} is increased enough to make emitter-to-base 1 voltage V_{EB} equal to $V_D + \eta V_{BB}$, the UJT triggers (goes from the off state to the on state). This value of V_{EB} is called *peak point voltage*, V_p. Thus

$$V_p = V_D + \eta V_{BB} \qquad (14\text{-}4)$$

Once $V_{EB} = V_p$, the forward-biased emitter diode injects holes into the r_{B1} region and electrons are drawn into the same region from terminal B_1. The increase in free current carriers lowers the value of r_{B1}. The decrease in r_{B1} sets off the triggering action shown in Fig. 14-5.

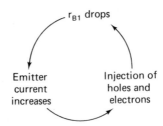

Figure 14-5 Diagram depicting UJT triggering action.

In Fig. 14-4, V_{EE} divides between R_E and r_{B1}. Since neither V_{EE} nor R_E changes, V_{EB} must decrease, because r_{B1} decreases. The collapse of r_{B1} is accompanied by an abrupt *increase* in emitter current and a *decrease* in emitter-to-base 1 voltage. Once r_{B1} begins to decrease, the UJT triggering action cannot stop until r_{B1} stops decreasing. The decreasing value of r_{B1} stops when saturation occurs at a value of r_{B1} called r_{sat}. Typical values of r_{sat} range between 5 and 25 Ω. The resistor R_E in Fig. 14-4 limits the emitter current after the UJT is triggered, to prevent destruction of the device. The above description of UJT action can be illustrated by current–voltage characteristics.

14-2 Current-Voltage Characteristics of a UJT

Figure 14-6 is a plot of emitter current I_E versus emitter-to-base 1 V_{EB}, for a given value of V_{BB}. This is the most common of the UJT characteristic curves. The off state is from the origin to point P, the on state is from point V to point S. A minimum value of emitter current is required at peak point P to start UJT triggering action. Manufacturer's data sheets give this value as *peak-point current* I_p, and typical values are 0.5 to 50 μA. This value is so small that it appears to be plotted on the horizontal axis. Between peak point P and valley point V, UJT triggering action is taking place and cannot be stopped. For this reason this path is called an *unstable state*.

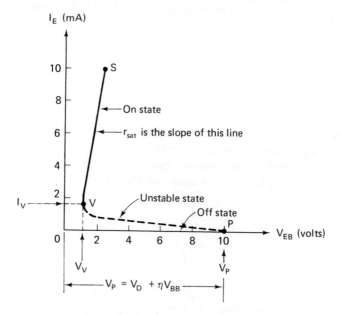

Figure 14-6 I_E–V_{EB} characteristics.

The value of emitter voltage at point P is peak-point voltage V_p. It is the voltage needed to trigger the UJT from off state to an on state and is given by Eq. (14-4). Point V is called the *valley point*. Manufacturer's data sheets give *valley current* I_V. Typical values of I_V range from 1 to 10 mA, depending on the UJT. *Valley voltage* V_V is not given on data sheets because V_V depends on the circuit. However, V_V will usually be between 1 and 3 V.

The UJT is turned off (that is, goes from the on state to the off state) by reducing the emitter current below I_V. When a UJT turns off, the voltage from emitter to base 1 drops to V_V, not V_p.

14-3 Relaxation Oscillator Using a UJT

The circuit of Fig. 14-7 is a timer or oscillator circuit. The frequency of oscillation depends on the values of R_E and C_E. When the switch in Fig. 14-7(a) is closed, the capacitor charges toward V_{BB} at a rate determined by R_E and C_E. During the time when the capacitor is charging, the emitter diode of the UJT is reverse-biased and therefore is essentially disconnected from the circuit. When voltage across the capacitor reaches peak-point voltage V_p, the UJT triggers from its off state to its on state. The capacitor then discharges through the UJT and R_L. This sudden increase in current through R_L causes a spike voltage across R_L, as shown in Fig. 14-7(c). Typically, values of R_L range from 10 to 100 Ω.

The capacitor voltage wave form is shown in Fig. 14-7(b) and illustrates how the capacitor discharges from V_p to V_V. At this time the UJT is turned off and capacitor begins to recharge.

The approximate frequency of oscillation is given by

$$f = \frac{1}{T} = \frac{1}{R_E C_E} \qquad (14\text{-}5)$$

If R_E is a 0- 100-kΩ variable resistor, the frequency can be varied. However, there is a minimum and maximum value of R_E to guarantee that the UJT continues to oscillate. The approximate practical minimum value of R_E is

$$R_{E\min} \approx 3\frac{V_{BB}}{I_V} \qquad (14\text{-}6)$$

where I_V is the valley current and given on manufacturer's data sheets. The maximum value of R_E is found by

$$R_{E\max} = \frac{V_{BB} - V_p}{I_p} \qquad (14\text{-}7)$$

where I_p is the peak-point current and also given on the data sheets.

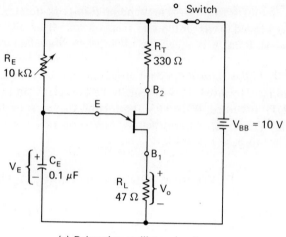

(a) Relaxation oscillator circuit

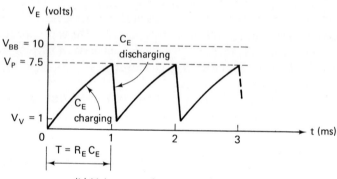

(b) Voltage waveform across the capacitor

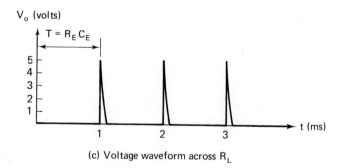

(c) Voltage waveform across R_L

Figure 14-7 Unijunction transistor relaxation oscillator and voltage wave forms.

If the UJT undergoes a temperature change r_{BB}, η and V_D change. This will change V_p and oscillator frequency. It has been found by experiment that resistor R_T in series with B_2 can stabilize V_p against temperature change. Values of R_T range from 100 Ω to 3.3 kΩ.

The simplest way to measure the UJT electrical characteristics η, V_p, V_V, I_V, and I_p is to build an oscillator and measure the capacitor voltage wave form with an oscilloscope. The wave shapes should be like those of Fig. 14-7(b) and (c).

Example 14-2: Use Fig. 14-7(b) to find (a) intrinsic standoff ratio and (b) frequency. Assume that $V_D = 0.5$ V.
Solution:

(a) From Fig. 14-7(b) $V_p = 7.5$ V; rearranging Eq. (14-4),

$$\eta = \frac{V_p - V_D}{V_{BB}} = \frac{7.5 \text{ V} - 0.5 \text{ V}}{10 \text{ V}} = 0.7$$

(b) From Fig. 14-7(b), the time between peaks is 1 ms and equals the period T. Then

$$f = \frac{1}{T} = \frac{1}{1 \text{ ms}} = 1 \text{ kHz}$$

As a check, using Eq. (14-5),

$$f = \frac{1}{R_E C_E} = \frac{1}{(10 \text{ k}\Omega)(0.1 \text{ }\mu\text{F})} = 1 \text{ kHz}$$

The output voltage spike, V_o, of Fig. 14-7(c) may be used to drive a relay or other electronic device. V_E in Fig. 14-7(b) may be used as a sawtooth or ramp voltage generator.

14-4 Constant-Current Charging UJT Oscillator

Figure 14-7(b) shows that the voltage across the capacitor rises exponentially. Figure 14-8(a) is a modification of Fig. 14-7(a). This new circuit replaces R_E with a constant-current charging circuit using a BJT and allows C to be charged at a constant rate. In Fig. 14-8(a), I_{B2} flows through and forward-biases the three silicon diodes. The total voltage across the diodes is approximately 3×0.6 V $= 1.8$ V. This voltage forward-biases the base–emitter junction of the BJT Q_2 and produces a voltage drop across R_E. Since the voltage drop across the emitter–base junction is about 0.6 V, the voltage across R_E is 1.8 V $-$ 0.6 V $= 1.2$ V. The emitter current and collector

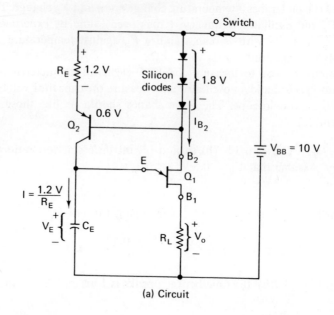

(a) Circuit

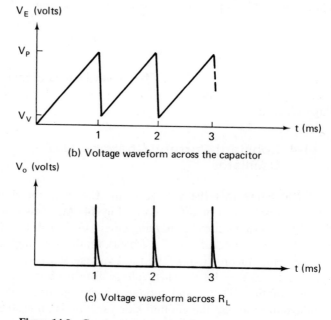

(b) Voltage waveform across the capacitor

(c) Voltage waveform across R_L

Figure 14-8 Constant-current charging circuit and wave forms.

current of a BJT are about equal. Therefore, the current charging the capacitor is

$$I = \frac{1.2 \text{ V}}{R_E} \tag{14-8}$$

The period of oscillation for the circuit of Fig. 14-8(a) is given by

$$T = \frac{(V_p - V_V)C_E}{I} \tag{14-9}$$

where V_p is the peak-point voltage, V_V is the valley voltage, C_E is the value of the capacitor, and I is given by Eq. (14-8). Figure 14-8(b) is the voltage across the capacitor and is linear during charging intervals. Compare this wave form with that of Fig. 14-7(b) and note that the charging interval is no longer exponential. The output wave form shown in Fig. 14-8(c) is similar to that of Fig. 14-7(c).

14-5 Programmable Unijunction Transistor

The programmable unijunction transistor (PUT) is not a single-junction semiconductor device like the UJT but is constructed from four alternating layers of p-type and n-type material, as shown in Fig. 14-9(a). The trigger voltage of the UJT given by Eq. (14-4) depends on the *intrinsic standoff ratio*, η, which is fixed by the manufacturing of the device. The standoff ratio of the PUT is established by an external voltage, resistor, or resistor network. Thus the word *programmable* signifies that we can program the peak-point voltage by selection of resistors.

Although the PUT is a four-layer device, only three terminals are brought

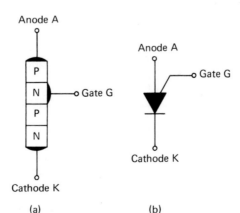

Figure 14-9 Structure and circuit symbol of a PUT.

(a)

(b)

out. These terminals are A for *anode*, K for *cathode*, and G for *gate*. The circuit symbol for the PUT is given in Fig. 14-9(b).

14-5.1 Current–Voltage Characteristics of a PUT. Figure 14-10 is a plot of anode current–voltage characteristics of a PUT. They are similar to the emitter characteristics of the UJT given in Fig. 14-6. The current axis in Fig. 14-10 is anode current I_A. This is the current that flows from anode A to cathode K. The voltage axis V_{AK} is the voltage measured from anode to cathode. A PUT, like a UJT, has two operating states, an off state and an on state. In switching from one state to the other it is necessary to go through an unstable state. You will remember from the UJT discussion that this third state is called an unstable state, because the device will not stay in this region. When the device triggers, it passes through the unstable state in a few microseconds.

The PUT triggers from the off state to the on state when the voltage from anode to cathode reaches peak-point voltage V_p, where

$$V_p \approx 0.5 \text{ V} + \eta V_{BB} \tag{14-10}$$

14-5.2 Program Voltage. Unlike the UJT, in which η is fixed, η of the PUT depends on the circuit in which the device is used. The term ηV_{BB} in Eq. (14-10) represents a *program voltage* and is symbolized by V_s, where

$$V_s = \eta V_{BB} \tag{14-11a}$$

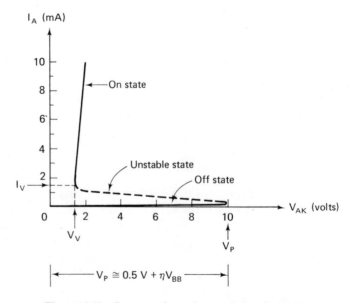

Figure 14-10 Current–voltage characteristics of a PUT.

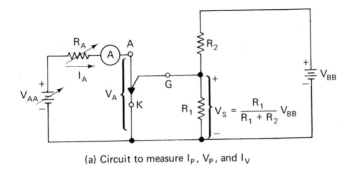

(a) Circuit to measure I_P, V_P, and I_V

(b) Model of the program voltage V_S

Figure 14-11 Resistors R_1 and R_2 in (a) allow selection of program voltage V_s in (b).

As shown in Fig. 14-11(a), program resistors R_1 and R_2 set the value of program voltage V_s according to the voltage divider law

$$V_s = \frac{R_1}{R_1 + R_2} V_{BB} \qquad \text{where } \eta = \frac{R_1}{R_1 + R_2} \qquad (14\text{-}11b)$$

As shown in the equivalent-circuit model of Fig. 14-11(b), program voltage is applied to the gate. The equivalent series gate resistance R_s of R_1 and R_2 is expressed by

$$R_s = R_1 \| R_2 = \frac{R_1 R_2}{R_1 + R_2} \qquad (14\text{-}11c)$$

14-5.3 Peak, Valley, and Holding Currents. When anode voltage V_A in Fig. 14-11(a) is increased to reach peak-point voltage V_p, a minimum value of anode current must flow to initiate triggering of the PUT. This minimum anode trigger current is called peak-point current I_p. Valley current I_V depends on equivalent gate series resistance R_s. Typical manufacturer's data show that I_V decreases from 100 μA with $R_s = 10$ kΩ to $I_V = 10$ μA at $R_s = 1$ MΩ. The PUT will remain in its on state even if the program voltage is removed from the gate. By reducing V_{AA} or increasing R_A in Fig. 14-11, anode current can be reduced to a minimum value, called *holding current*, where the PUT abruptly switches to its off state.

14-6 PUT Relaxation Oscillator

14-6.1 Capacitor Charge Time. In the relaxation oscillator of Fig. 14-12, R_1, R_2, and V_{BB} set gate program voltage V_s. When the switch is closed, current through timing resistor R_A charges capacitor C. Since C is connected across anode and cathode and no cathode current flows from the off PUT, anode voltage V_A equals capacitor voltage. The time, T, for capacitor C to charge from V_V to peak-point voltage V_p is given by

$$T = R_A C \ln \frac{V_{BB} - V_V}{V_{BB} - V_p} \tag{14-12}$$

In this case the initial capacitor charge was zero volts, so substitute 0 for V_V in Eq. (14-12). [In Eq. (14-12) ln represents the logarithm to the base e or natural logarithm (not the logarithm to the base 10).]

14-6.2 Minimum and Maximum Anode Resistance. Once V_A reaches V_p, the PUT will trigger on only if R_A is small enough to allow the required peak-point current I_p to flow. This maximum value of R is

$$R_{A\,\text{max}} = \frac{V_{BB} - V_p}{I_p} \tag{14-13a}$$

If R_A is less than $R_{A\,\text{max}}$, the PUT will trigger into its on state. Then one of two possibilities can occur. If anode current is greater than valley current I_V, the PUT stays in its stable on state until I_A is reduced below I_V. However, if

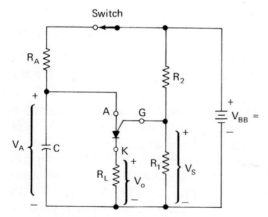

Figure 14-12 Relaxation oscillator using a PUT.

R_A is made large enough to keep I_A *below* I_V, the capacitor will discharge its energy through the PUT and develop an output voltage spike across R_L. Anode voltage and capacitor voltage will drop to V_V, and since the PUT cannot remain in its unstable state, it abruptly turns off. C will begin to recharge and the charge–discharge cycle repeats, to yield an oscillator. To maintain oscillation, R_A must be greater than a minimum value given by

$$R_{A\,min} = \frac{V_{BB} - V_V}{I_V} \tag{14-13b}$$

Example 14-3: In Fig. 14-12, $R_1 = 16\,k\Omega$, $R_2 = 27\,k\Omega$, and $V_B = 20\,V$. Find (a) program voltage V_s, (b) peak-point voltage V_p, and (c) gate series resistance R_G.
Solution:

(a) From Eq. (14-11b), $V_s = \dfrac{27\,k\Omega}{16\,k\Omega + 27\,k\Omega}(20\,V) = 12.5\,V$

(b) From Eq. (14-10), $V_p = 0.5 + V_s = 0.5 + 12.5 = 13\,V$

(c) From Eq. (14-11c), $R_s = 16\,k\Omega \| 27\,k\Omega = 10\,k\Omega$.

Example 14-4: In the oscillator of Example 14-3, pulses stop when R_A is increased to 7 MΩ. This is a measurement of $R_{A\,max}$. Calculate the value of peak-point current I_p.
Solution: From Eq. (14-13a),

$$R_{A\,max} = \frac{V_{BB} - V_p}{I_p} \quad \text{or} \quad I_p = \frac{V_{BB} - V_p}{R_{A\,max}} = \frac{(20 - 13)\,V}{7\,M\Omega} = 1\,\mu A$$

This example shows a simple technique for the measurement of I_p.

Example 14-5: If valley current I_V is specified as 100 μA for the PUT in Examples 14-5 and 14-4 calculate $R_{A\,min}$. Assume that V_V is 1 V.
Solution: Eq. (14-13b),

$$R_{A\,min} = \frac{(20 - 1)\,V}{100 \times 10^{-6}\,A} = 190\,k\Omega$$

For margin the actual value of $R_{A\,min}$ should be more than twice as large as the calculated value of $R_{A\,min}$.

14-6.3 Frequency of Oscillation. The time for C in Fig. 14-12 to charge from V_V to V_p represents one complete cycle of oscillation. Usually the discharge time for C is negligible with respect to its charge time, so we can express oscillator frequency f by

$$f = \frac{1}{T} \qquad (14\text{-}14)$$

where T is found from Eq. (14-12).

Example 14-6: Given $V_{BB} = 21$ V, $V_V = 1$ V, $V_p = 11$ V, $R_A = 1$ MΩ, and $C = 1$ μF in Fig. 14-12, find the (a) charge time and (b) frequency of oscillation.
Solution:

(a) From Eq. (14-12),

$$T = (10^6 \text{ Ω})(1 \times 10^{-6} \text{ F}) \ln \frac{21 - 1}{21 - 11} = 1 \ln 2 = 0.69 \text{ s}$$

(b) From Eq. (14-14),

$$f = \frac{1}{0.69} = 1.45 \text{ Hz}$$

Problems

14-1 How is the interbase resistance of a UJT measured?

14-2 Approximately what resistance value will be measured for interbase resistance?

14-3 The intrinsic standoff ratio of a UJT equals 0.7 and $r_{BB} = 10$ kΩ. Find (a) r_{B1}; (b) r_{B2}.

14-4 If $V_{BB} = 20$ V and $n = 0.7$, find the voltage across r_{B1}.

14-5 Find the peak-point voltage V_p for Problem 14-4.

14-6 The UJT of Fig. 14-6 has $n = 0.68$. Find V_{BB}.

14-7 In Example 14-2 substitute another UJT with $n = 0.6$. To what voltage will the capacitor charge?

14-8 R_E is changed to 50 kΩ in Fig. 14-7. Find the new frequency of oscillation.

14-9 What oscillator frequency results in Fig. 14-7 when C is changed to 0.01 μF?

14-10 Find R_E in Fig. 14-7 to obtain an oscillator frequency of 500 Hz.

14-11 Choose R_E in Fig. 14-8 for $I = 100$ μA.

14-12 With $R_E = 12$ kΩ in Fig. 14-8 and replacing the three silicon diodes with four silicon diodes, find I.

14-13 Find period T and frequency of oscillation f for Fig. 14-8 if $I_V = 150$ μA, $C_E = 0.1$ μF, $V_p = 7.5$ V, and $V_V = 1.5$ V.

14-14 If I_V is changed to 100 μA in Problem 14-13, find the new frequency of oscillation.

14-15 In Fig. 14-11, $R_1 = R_2 = 20$ kΩ and $V_{BB} = 20$ V. Find (a) η; (b) R_s; (c) V_s; (d) V_p.

14-16 Find $R_{A\,max}$ and $R_{A\,min}$ in Fig. 14-11 for $V_{BB} = 20$ V, $V_p = 13$ V, $I_p = 2$ μA, $I_V = 200$ μA, and $V_V \simeq 0$ V.

14-17 Find the frequency of oscillation in Example 14-6 if R_A is set at its upper limit of 7 MΩ.

14-18 Find the frequency of oscillation in Example 14-6 if R_A is set at its lower limit of 190 kΩ.

14-19 If $C = 0.01$ μF in Example 14-6, does the frequency of oscillation decrease or increase by a factor of 100?

15

Silicon-Controlled Rectifiers and Triacs

15-0 Introduction

Multilayer semiconductor devices that are intended primarily as on–off switches are classified as part of the thyristor family. There are two general groupings of thyristors: *unidirectional* or *unilateral* thyristors, which conduct current through their switch terminals in one direction only; and *bidirectional* or *bilateral* thyristors, which conduct current in both directions. The silicon-controlled rectifier (SCR) and triac are the most popular examples of uni- and bidirectional thyristors, respectively.

When a thyristor blocks current flow, it acts like an open switch. When a thyristor turns on or conducts current, it acts like a closed switch. The conducting pair of terminals are analogous to switch terminals and can conduct heavy currents, of the order of amperes. A third, *gate* terminal allows a very small gate current (milliamperes) to turn the thyristor on. So a tiny gate current can control large load currents.

If a sinusoidal line voltage is applied to a load in series with the thyristor, the gate can turn on the thyristor at any part of the sine wave. All the line voltage, or a fraction of the line voltage, can thus be applied to the load during the time the thyristor is on. When the thyristor is on, it consumes very little power. So a thyristor's main advantage is that it can control large amounts of power with very little wasted power. Contrast this advantage with use of a series resistor to limit load current and power. As power to the load is reduced by increasing series resistance, the resistor absorbs and wastes more power. Operation of both SCR and triac will be studied beginning with the SCR.

258

15-1 Silicon-Controlled Rectifier

Figure 15-1(a) shows a simplified model of a silicon-controlled rectifier. It is a four-layer device with three external leads—*anode, A*; *cathode, K*; and *gate, G*. The circuit symbol is given in Fig. 15-1(b). The arrowhead indicates the direction of conventional current, from anode to cathode. Since current only flows in one direction, the SCR is classified as a unidirectional device. The SCR has two operating conditions, an off state and an on state. In the off state the SCR's anode and cathode act like an open switch; in the on state they act like a closed switch. The conventional method of turning an SCR on is by connecting a small control voltage between gate and cathode. Once the SCR is turned on, the gate has no further control over the device. In fact, the voltage between gate and cathode that was used to turn the device on may be removed and the SCR will stay on. To turn the SCR off, the anode-to-cathode current must be reduced below a specified minimum value.

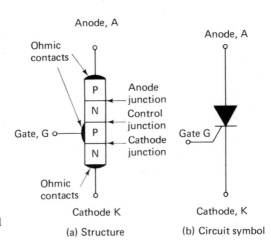

Figure 15-1 Structure and symbol of a silicon-controlled rectifier.

(a) Structure (b) Circuit symbol

15-2 Current–Voltage Characteristics of an SCR

Typical forward and reverse characteristics of an SCR are given in Fig. 15-2 for the special case where the gate voltage is zero. When the anode is made positive with respect to the cathode, as shown by V_{AKF} in Fig. 15-2, the control junction is reverse-biased and there is no flow of forward current I_{AKF}. When forward anode-to-cathode voltage V_{AKF} is increased, no forward current flows from anode to cathode until forward breakover voltage, V_{FXM}, is abruptly reached. This voltage will trigger the SCR abruptly from the off state to the on state. When an SCR is on, forward current is limited by *R*

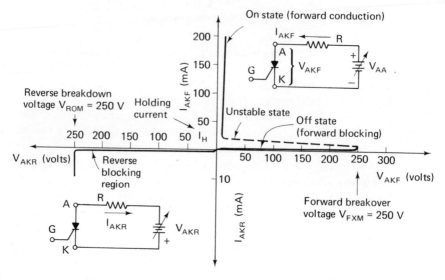

Figure 15-2 Forward and reverse characteristics of a silicon-controlled rectifier.

and V_{AKF}. Voltage drop across anode and cathode is approximately 1 V. The off state is also referred to as the *forward blocking state*, and the on state is also called the *forward conduction state*. In practical applications the SCR is *not* triggered by increasing V_{AKF} beyond the breakover voltage. Rather, V_{AKF} is kept *less* than the breakover voltage and a trigger voltage is applied between gate and cathode. Gate triggering and other methods of triggering the SCR are discussed in the following sections. Once the SCR is on, the only way to turn it off is by reducing the anode-to-cathode current below a small value, called the *holding current*, I_H. The holding current depends on the SCR but is usually between 5 and 30 mA. When a positive voltage is applied to the cathode and a negative voltage to the anode, both anode and cathode junctions are reverse-biased. The SCR is off because of this reverse blocking voltage V_{AKF}. Figure 15-2 shows that in the reverse blocking region, only a negligible leakage current flows. The SCR breaks down like a conventional diode at a value of voltage called the *reverse breakdown voltage*, V_{ROM}. Normally, V_{ROM} is approximately equal to V_{FXM}.

15-3 Conventional Method of Triggering an SCR

To change the SCR from the off state to the on state it is necessary to supply free holes and electrons to the cathode junction. The most common method of accomplishing this is by applying a forward-bias control voltage

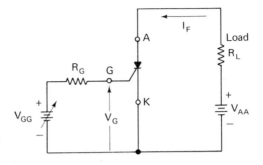

Figure 15-3 Circuit to obtain minimum gate trigger voltage for a specified V_{AA}.

between gate and cathode. In Fig. 15-3, V_{GG} is increased to raise gate-to-cathode voltage V_G to a value large enough to forward-bias the cathode junction and turn the SCR on. This value of V_G is called *gate trigger voltage* V_{GT}. Since the gate-to-cathode junction is a *pn* junction, the values of V_{GT} needed to forward-bias this junction are in the order of 0.6 V. When the SCR triggers on, anode current I_A changes from practically zero to a value determined by V_{AA} and R_L. The voltage between anode and cathode of an "on" SCR is about 1 V.

15-4 Other Methods of Triggering an SCR

15-4.1 Forward Breakover. Voltage V_{AA} in Fig. 15-3 can be increased to a value where the reverse bias on the control junction causes an increase in leakage current. This leakage current supplies holes and electrons to the cathode junction and will trigger the SCR just as it did when a gate voltage was applied. Manufacturer's data sheets give a value for *repetitive forward blocking voltage* V_{DRM} or V_{FXM}. This voltage rating is a minimum value. Once this voltage is exceeded, the SCR turns on and will remain on even with the gate open. SCRs may be purchased with V_{FXM} values between 15 and 800 V. Triggering an SCR by exceeding V_{FXM} will not harm the SCR, but the false turn on can cause harm. For example, the false turn on could start an elevator motor, with the doors open.

15-4.2 Rate Effect. If an external voltage such as V_{AA} in Fig. 15-3 is suddenly connected to the SCR, the anode-to-cathode voltage rises very fast. This rapid increase in anode-to-cathode voltage may cause the SCR to turn on, because a charging current flows due to the capacitance associated with *pn* junctions. This charging current flows into the cathode junction, causing a false turn on. Manufacturers specify the maximum rate of change of voltage that can safely be applied from anode to cathode. For example, a manufacturer's data sheet gives this minimum value as 10 V/μs. If V_{AA} is set at 100 V and a switch is closed connecting it to the SCR in about 4 μs, the rate of change of voltage is 100 V/4 μs = 25 V/μs. This would cause the SCR to turn on. A series RC network is connected between anode and cathode to reduce the rate of change of V_{AK}. Typical values for R are 10 Ω; for C, 0.05 μF.

15-4.3 Temperature. It was seen from Section 3-1.2 that leakage current increases when temperature increases. For the SCR, this leakage current flows through the cathode junction; and if the temperature of the SCR becomes high enough, it will turn on.

15-4.4 Radiant Energy. Radiant energy striking semiconductor material will break covalent bonds to create holes and electrons. The freed holes and electrons act as a gate current in the cathode junction and will turn on the SCR. This principle is used for the operation of the *light-activated silicon-controlled rectifier* (LASCR), because light is radiant energy.

15-5 UJT–SCR Application

An output voltage pulse from base 2 of a UJT can be applied between gate and cathode of an SCR. If the UJT pulse voltage exceeds V_{GT}, the SCR will turn on. The SCR is used to control large values of current through a load when triggered on. Figure 15-4 shows how a UJT and SCR are connected together. When the switch is closed, C_E charges at a rate determined by R_E and C_E. When the voltage across C_E reaches V_p [the peak-point voltage of the UJT given by Eq. (14-4)], the UJT turns on. The resulting output voltage pulse V_{op}, across R_L, is connected to the gate of the SCR and the SCR turns on. Remember that once the SCR turns on, it remains on until gate voltage is removed and anode-to-cathode current goes below the holding current.

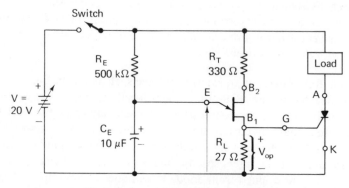

Figure 15-4 UJT–SCR timing circuit.

Some care must be taken in the selection of R_L. R_L must be small so that the voltage drop across it, before the UJT fires, is less than V_{GT} (gate turn-on voltage). But R_L must also be large enough so that when the UJT does fire, V_{op} exceeds V_{GT}. A good choice for R_L is between 10 and 47 Ω. The UJT–SCR combination is one method for controlling an alarm system with a specified time delay. This delay is accomplished by selecting the proper values of R_E and C_E. The values of R_E and C_E in Fig. 15-4 give a 5-s delay after the switch is closed.

15-6 Phase Control

Phase control is a process whereby an ac voltage is connected to the load for only part of each cycle. An SCR acts as a switch to make the connection because it can be rapidly switched on and off. Only when the SCR is on can current flow through the load. Since power is related to current ($P = I^2R$), the SCR controls power delivered to the load. This is an efficient and inexpensive method of controlling power to, for example, lamps, heaters, or motors.

Figure 15-5 illustrates three general types of phase-control circuits that use one SCR to control power. The wave form for load and SCR current, I_F, of the half-wave phase control circuit in Fig. 15-5a appears similar to the half-wave rectified wave form of Fig. 3-6. However, the SCR may not be conducting for the entire half-cycle. It will conduct only after the SCR is triggered on by the control circuit. The control circuit may be as simple as a switch. Upon closing the switch a small voltage is applied to the gate and the SCR is triggered on. If we insert, for the control, a UJT circuit in which the firing time can be adjusted, the SCR can be turned on at any time during the positive half-cycle. Negative half-cycles of the input voltage reverse-bias the SCR, and no current flows. The shaded portion of the output wave form is the time when the SCR is on, or conducting current and power is delivered to the load. Reducing the shaded portion by triggering later decreases power to the load. Increasing the shaded portion by earlier triggering increases power to the load. Maximum power is delivered to the load when the SCR is conducting for the entire positive half-cycle.

A more efficient use of the SCR is shown in Fig. 15-5(b) and (c). In these circuits a full-wave rectifier is used so that both positive and negative half-cycles of the input voltage can be used. However, location of the load becomes important. For dc loads, the load is placed in series with the SCR, as shown in Fig. 15-5(b). Figure 15-5(c) shows that for ac loads, the load should be placed on the line side of the circuit. The output or load wave forms are also shown. For the dc phase-control circuit, the load-current wave form is the same as that which flows through the SCR. The load-current wave form in

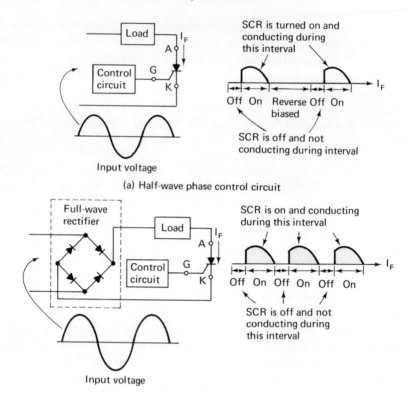

(a) Half-wave phase control circuit

(b) Full-wave dc phase control circuit

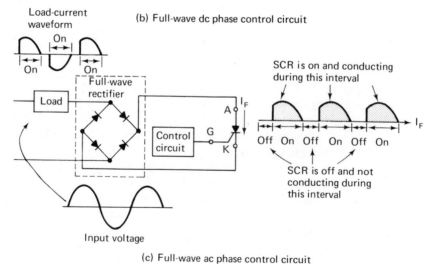

(c) Full-wave ac phase control circuit

Figure 15-5 Three types of phase-control circuits.

the ac phase-control circuit is an ac wave form. Current flows during both positive and negative half-cycles. Note that shaded portions correspond to periods when the SCR is on. The following two sections illustrate practical dc and ac phase-control circuits. In both cases a UJT circuit is used in the control circuit.

15-6.1 DC Phase Control. A variable-voltage dc control circuit is illustrated in Fig. 15-6. The four diodes form a full-wave rectifier (Section 3-3.2). The Zener diode guarantees that the UJT will be protected in case the SCR does not trigger on. R_E and C_E control the time when the UJT fires the SCR. The "on" portion of the full-wave rectified voltage across the load depends on R_E and C_E and may be selected for any value between 0 and 180°.

15-6.2 AC Phase Control. A modification of Fig. 15-6 is shown in Fig. 15-7. The load is moved to the line or input side of the full-wave rectifier. In

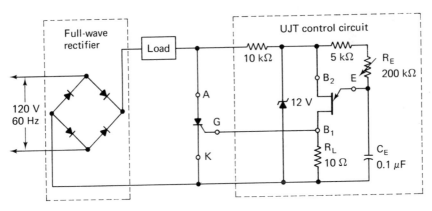

Figure 15-6 Dc phase-control circuit.

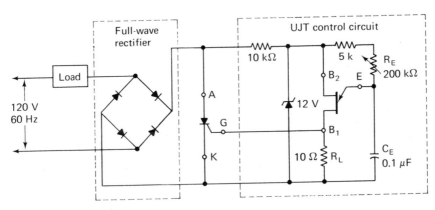

Figure 15-7 Ac phase-control circuit.

this position the load will conduct during positive and negative half-cycles but only when the SCR is on. The values of R_E and C_E determine when the UJT fires, which in turn triggers the SCR, thus controlling power to the load.

15-7 Triac

15-7.1 Introduction. Figure 15-8(a) shows that the *triac* is a three-terminal five-layer semiconductor device. The terminals of the triac are referred to as terminal T_1, terminal T_2, and gate G. All voltages are measured from T_1, and for this reason terminal T_1 is the reference terminal. Terminal T_2 is usually electrically connected to the case, to which a heat sink may easily be attached. The gate terminal is used to trigger the triac from an off condition to an on condition. Like the SCR, once the triac is triggered on, it remains

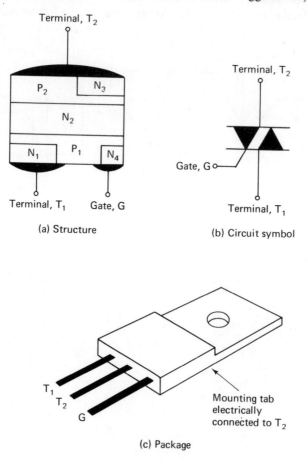

(a) Structure

(b) Circuit symbol

(c) Package

Figure 15-8 Structure, symbol, and package of a triac.

conducting, regardless of gate voltage, until current through T_1 and T_2 is reduced below a specified limit, called *holding current*.

Unlike the SCR, the triac conducts current in both directions and is therefore classified as a *bidirectional* or *bilateral device*. Figure 15-8(b) is the circuit symbol for the triac. The arrowheads point in the direction of conventional current flow, indicating that current flows either from T_2 to T_1 or from T_1 to T_2. Since a triac is a bidirectional device, its applications are as a power switch for alternating voltages and current.

15-7.2 Current–Voltage Characteristics of a Triac. Figure 15-9 shows the forward and reverse characteristics of a triac. Both characteristics are the same and are similar to the forward characteristics of an SCR. Forward breakover voltage V_{DROM} or reverse breakover voltage V_{DROM}, is given on data sheets. This voltage triggers the triac from the off state to the on state with no signal applied to the gate. As with the SCR, the conventional method of triggering the triac is to apply a voltage from T_2 to T_1, less than V_{DROM}, and then apply a gate voltage between gate and T_1. *Positive conduction* refers to current flow through the triac from T_2 to T_1; while *negative conduction* refers to current direction from T_1 to T_2. In both positive and negative conduction cycles, the triac switches from the on state to the off

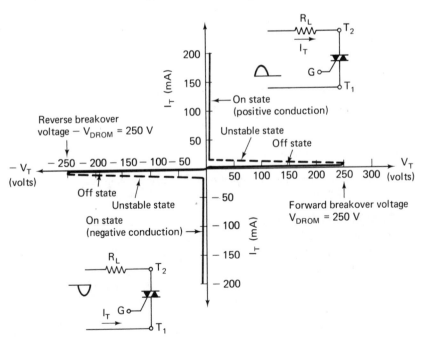

Figure 15-9 Forward and reverse characteristics of a triac.

state when the T_1–T_2 current becomes less than the holding current I_H. The characteristics show that once the device turns on, in either direction, the voltage across terminals T_1 and T_2 is approximately 1 V. Also, when the device is on, terminal current I_T must be limited by load resistor R_L.

15-7.3 Triggering the Triac. Current flow T_2 to T_1 within the triac is referred to as *positive* or *forward conduction*. Under forward conduction, terminal T_2 is positive with respect to T_1 by about 1 V. Figure 15-10(a) shows positive conduction I_T flowing through the equivalent SCR of $P_2N_2P_1N_1$. The *I–V* characteristic of Fig. 15-9 is that of an SCR in the forward direction. Positive conduction is also called *mode I operation*.

Negative or *reverse conduction* describes current flow from T_1 to T_2. Under this conduction, terminal T_1 is positive with respect to T_2. The *I–V* characteristic for negative conduction is also similar to that of an SCR. The equivalent SCR is $P_1N_2P_2N_3$, as shown in Fig. 15-10(b). Negative conduction is also referred to as *mode III operation*.

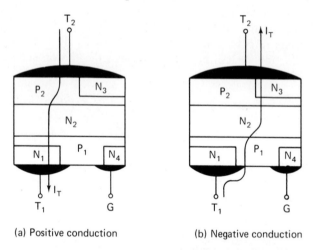

(a) Positive conduction   (b) Negative conduction

Figure 15-10 Direction of current through a triac for both positive and negative conduction.

The SCR can only be triggered by a positive gate voltage. However, the triac can be triggered by making the gate either positive or negative with respect to T_1. When the gate is positive with respect to T_1, the P_1N_1 junction is forward-biased and the triac is triggered on. Triggering the triac with negative gate pulses forward-biases the P_1N_4. For best performance, most applications use negative gate pulses.

15-7.4 Switching Applications. The switch in Fig. 15-11 controls only the tiny gate current and can be inexpensive. On point 1 the triac remains off.

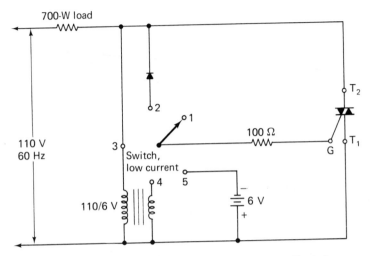

Figure 15-11   Methods of using the triac as an on–off switch.

On point 2 only negative voltage is applied to the gate, so the triac can only turn on during negative half-cycles for half-wave load power. When line voltage rises to about -1 V, the triac negative-conducts. The voltage across T_2 and T_1 stays at the on voltage of 1 V and protects the diode and its own gate. As line voltage goes through its negative half-cycle load, triac current goes through a maximum, then reduces below I_H to turn off the triac.

When the switch is thrown to point 3, the triac conducts fully during both negative and positive half-cycles and turns off at the end of each half-cycle.

Switch points 4 and 5 show how a remote, low-power, ac and dc voltage, respectively, can turn the triac on or off. Milliamperes of gate current from the 6-V supply control amperes of load current from the 110-V line.

15-8 Diac and Neon Lamp Trigger

There are other triggering devices in addition to the UJT. The *diac* is a symmetrical trigger diode that conducts in either direction. Its symbol is shown in Fig. 15-12(b) and *I–V* characteristic in Fig. 15-12(a). The neon lamp of Fig. 15-12(c) has similar characteristics. Aside from their differences in appearance, cost, and construction, they differ in triggering voltage, V_p, and valley voltage, V_v. Trigger voltage for the neon lamp is typically 90 V and is typically 20 V for the diac.

Both diac and lamp are used to sense voltage across a capacitor; and, as long as capacitor voltage is below V_p, the voltage is isolated from a gate control. When capacitor voltage exceeds V_p, the diac or lamp breaks down and the terminal voltage drops to V_v. The difference in voltage between V_p and

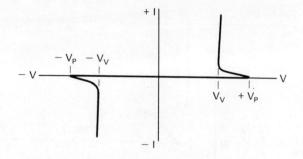

(a) Current-voltage characteristics

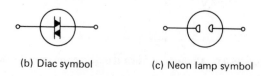

(b) Diac symbol   (c) Neon lamp symbol

Figure 15-12 Current–voltage characteristics in (a) are similar for diac (b) and neon lamp (c).

V_v is coupled as a pulse to a gate (SCR or triac) to turn it on. A diac is selected to illustrate this action in the next section.

15-9 Diac-Triac Light Control

In Fig. 15-13 resistor R can be adjusted to change the rate at which capacitor C charges. As voltage across C increases, no current flows through the

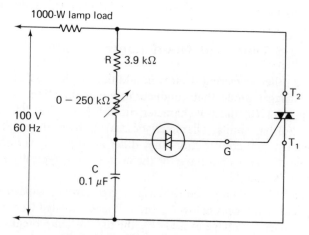

Figure 15-13 Diac–triac lamp dimmer.

diac until capacitor voltage reaches V_p. As shown in Fig. 15-12(a), the diac abruptly triggers on and conducts current into the gate terminal to turn on the triac. The triac then completes the circuit to the lamp load until triac current drops below I_H.

Once the triac is turned on, its terminals T_2 and T_1 place a short across R and C and allow C to discharge. When capacitor voltage drops below V_v, the diac turns off, which removes gate current from the triac.

Since resistor R can vary how swiftly the capacitor voltage rises, R controls when the diac will fire. Reducing R fires the diac and triac earlier in each half-cycle, and the load conducts more current to develop more power. For a lamp load the lamp would become brighter. Increasing R fires the triac later in each cycle, thus dimming the lamp.

Problems

15-1 In Fig. 15-2, $V_{AA} = 300$ V of forward bias, forward breakover voltage is 200 V, and $R = 1500\ \Omega$. Find forward current I_{AKF}.

15-2 If $V_{AA} = 100$ V of forward bias and $V_{FXM} = 200$ V, find I_{AKF}.

15-3 Approximately what value of V_{GG} is required in Fig. 15-3 to turn the SCR on?

15-4 In Fig. 15-3, $R_L = 100\ \Omega$ and $V_{AA} = 100$ V. Find the anode current when the SCR is on.

15-5 Why is a 10-Ω resistor and 0.05-μF capacitor connected in series across the anode and cathode of an SCR?

15-6 Why is no power delivered to the loads in Fig. 15-5 when the SCR is off?

15-7 Assume that one diode becomes an open circuit in Fig. 15-5(b). What change would occur in the SCR current?

15-8 One half-cycle of a 60-Hz ac voltage has a time interval of 8.3 ms. Show that R_E in Fig. 15-6 must be 40 kΩ to fire the SCR 4 ms after the beginning of each half-cycle? [*Hint:* Use Eq. (14-5).]

15-9 Assume that the UJT in Fig. 15-6 does not fire. What protective action is performed by the Zener?

15-10 When the SCR is on in Fig. 15-6, is the Zener on?

15-11 What is the difference in gate-triggering voltage between an SCR and a triac?

15-12 Does the triac in Fig. 15-11 experience positive or negative conduction when the switch is on point 2?

15-13 Why doesn't the gate burn out in Fig. 15-11 when the switch is on point 3?

15-14 Does increasing R in Fig. 15-13 dim or brighten the lamp?

16

Frequency, Power, and Temperature Limitations

16-0 Introduction

Once the basic operation of a device is understood, it is natural to investigate its limitations and also the limitations of the circuit in which it is connected. Two of these limitations, which are of primary concern to engineers and technicians, are frequency and power. Until now the signal input frequency has been set at approximately 1 kHz, which is called *midfrequency*. In this chapter we shall investigate what happens to the output voltage when the input frequency is decreased below 1 kHz and when it is increased above 1 kHz. The other prime performance limitation is heat. Or, to put it another way: How much power will the device withstand? This limit is established by the temperature range given by manufacturers on their data sheets. The first six sections deal with frequency limitations; Sections 16-7 and 16-8 treat temperature and power limitations.

16-1 Bandwidth and Cutoff Frequencies

16-1.1 Measurement Circuit. The common-emitter amplifier in Fig. 16-1(a) is used to show that output voltage, V_o, decreases as the oscillator's frequency, f, is increased. However, to compare a measurement of V_o at one frequency with that taken at another frequency, the *amplitude* of input signal voltage E_i must be kept constant. E_i represents a laboratory-type variable-frequency oscillator. An ac voltmeter is connected across the oscillator's terminals to monitor the amplitude of E_i, and the oscillator's volume control

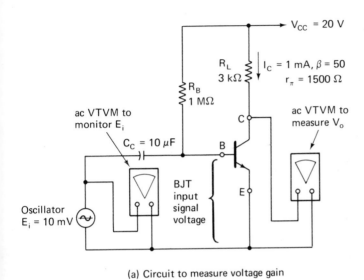

(a) Circuit to measure voltage gain
variation with frequency

V_o (volts)

V_o versus f for E_i = 10 mV

Bandwidth b = f_H − f_L

V_o at f_L = 0.707(V_o at M) = V_0 at f_H

E_i, frequency (hertz) ⟶

(b) Frequency-response curve

Figure 16-1 Frequency of E_i and amplitude of V_o are measured in (a) to yield the frequency-response curve in (b).

is adjusted to maintain E_i constant regardless of what changes may occur in the BJT circuit. Voltage gain (V_o/E_i) is 100 when the frequency control on the oscillator is set at 1000 Hz. Thus when the amplitude of E_i is set for 10 mV, V_0 equals 100×10 mV $= 1000$ mV $= 1.0$ V.

16-1.2 Measurement of Lower Cutoff Frequency. Suppose that the frequency of E_i is decreased but that the amplitude of E_i is held constant. V_o remains constant at 1.0 V until a frequency is reached at which V_o drops to 0.707 V, or 0.707 times its value at 1000 Hz. The value of this frequency is read from the oscillator's frequency dial and is recorded as *lower cutoff frequency* f_L. As frequency is reduced below f_L, V_o continues to decrease. The action of V_o is shown in the *frequency-response curve* of Fig. 16-1(b). It is called a frequency-response curve because it shows how V_o responds to changes in frequency (not amplitude) of E_i. Point M (stands for midfrequency) shows V_o at 1 kHz and point L shows V_o at f_L. Since E_i is held constant, voltage gain $V_o/E_i = A_V$ varies in the same manner as V_o. Thus the frequency-response curve could also be a plot of A_V rather than V_o against frequency.

16-1.3 Measurement of Upper Cutoff Frequency. Suppose that the frequency of E_i was reset to 1000 Hz, where V_o will be 1.0 V and voltage gain will be 100. Now, let the frequency of E_i increase (still maintaining an amplitude of E_i constant at 10 mV). V_o initially remains constant at 1.0 V. Then it begins decreasing until a particular value of high frequency is reached where V_o drops by a factor of 0.707 to 0.707 V. The value of this *high* or *upper cutoff frequency* is read from the oscillator's frequency dial and recorded as f_H. The action of V_o during this high-frequency range is shown by the frequency-response curve between points M and H in Fig. 16-1(b). As the frequency is increased above f_H, V_o continues to decrease.

The useful frequency range of an amplifier contains all the frequencies that lie between f_L and f_H. This frequency range is called *bandwidth* and is given the symbol BW. The bandwidth is calculated from

$$BW = f_H - f_L \qquad (16\text{-}1)$$

16-1.4 Summary. At both the *lower cutoff frequency*, f_L, and the *upper cutoff frequency*, f_H, output voltage, V_o, as well as voltage gain, A_v, are reduced to 0.707 times their value at 1000 Hz. At f_L and f_H,

$$V_o = 0.707 \times (V \text{ at } 1 \text{ kHz}) \qquad (16\text{-}2)$$

$$A_v = 0.707 \times (A_v \text{ at } 1 \text{ kHz}) \qquad (16\text{-}3)$$

The bandwidth, BW, is the frequency range from f_L to f_H.

Thus far we have not examined why the useful frequency range limits occur when V_o is reduced by 0.707. This issue is considered next.

16-2 Half-Power Frequencies

The power delivered to R_L when E_i is applied and set to 1000 Hz is shown in Fig. 16-2(a) to be

$$\text{output power at 1 kHz: } P_o = \frac{V_o^2}{R_L} \qquad (16\text{-}4)$$

At both lower and upper cutoff frequencies, output signal power is shown in Fig. 16-2(b) to be reduced by half, to

$$\text{output power at } f_L \text{ and } f_H: P_L = P_H = \frac{1}{2}\frac{V_o^2}{R_L} \qquad (16\text{-}5a)$$

By comparing Eqs. (16-4) and (16-5a), we conclude that power levels at both f_L and f_H are one-half of the power available at 1 kHz, or

$$P_L = P_H = \tfrac{1}{2}P_o \qquad (16\text{-}5b)$$

Equation (16-5b) defines the useful frequency range or bandwidth. When output power is reduced by more than one-half, it is considered to be no longer useful. This is the situation that occurs for frequencies below f_L and above f_H. Thus f_L and f_H may also be called *half-power*, or 0.707, *frequencies*, as well as *lower* and *upper cutoff frequencies*.

Example 16-1: The voltage gain of a circuit is 200 at 1 kHz. Find (a) the voltage gain at f_H, (b) the voltage gain at f_L, (c) V_o at 1 kHz if $E_i = 10$ mV, and (d) V_o at f_L and f_H.

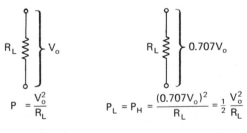

(a) Power at 1000 Hz (b) Power at f_L and f_H

Figure 16-2 Signal-power levels for points *L*, *M*, and *H* on the response curve of Fig. 16-1.

Solution:

(a) From Eq. (16-3), A_v at $f_H = 0.707 \times 200 = 140$

(b) From Eq. (16-3), A_v at $f_L = 0.707 \times 200 = 140$

(c) V_o at 1 kHz $= (A_v$ at 1 kHz$)E_i = 200 \times 10$ mV $= 2000$ mV $= 2.0$ V

(d) From Eq. (16-2), V_o at f_L and $f_H = 0.707 \times 2.0$ V $= 1.41$ V

Example 16-2: In Example 16-1, $f_H = 20,000$ Hz and $f_L = 10$ Hz. Find the bandwidth.

Solution: From Eq. (16-1), $B = f_H - f_L = (20,000 - 10)$ Hz $\simeq 20,000$ Hz. We conclude that, if f_L is very small with respect to f_H, the bandwidth BW is approximately equal to f_H.

Example 16-3: If R_L is 1000 Ω in Example 16-1, find the output power at (a) 1000 Hz and (b) lower and upper cutoff frequencies.

Solution:

(a) From Eq. (16-4), $P_o = \dfrac{(2.0\ V)^2}{1000\ \Omega} = 4$ mW

(b) From Eq. (16-5b), $P_L = P_H = \frac{1}{2}(4$ mW$) = 2$ mW

We next look for the circuit or BJT elements that cause V_o to vary with frequency. This problem will be analyzed in two parts, first at the lower cutoff frequency and then at the upper cutoff frequency.

16-3 Determining Lower Cutoff Frequency

16-3.1 Action of Coupling Capacitor. At 1 kHz, the reactance of the coupling capacitor, C_C, in Fig. 16-1(a) is found from

$$X_C = \frac{1}{2\pi f C_C} = \frac{1}{2\pi(1000)(10 \times 10^{-6})} = 15.9\ \Omega$$

By drawing the ac model in Fig. 16-3 we see that X_C is negligible with respect to R_{in}. Note that $R_{\text{in}} = R_B \| r_\pi = 1$ M$\Omega \| 1500\ \Omega \simeq 1500\ \Omega$. Thus, all of E_i appears across the BJT's base and emitter terminals to be amplified and yield $V_o = A_v E_i = 100\ E_i$.

Now let the frequency of E_i be reduced. The reactance of C_C must increase as frequency is lowered. Figure 16-3(a) shows that E_i must divide between X_C and R_{in}. Thus at lower frequencies V will be less than E_i. Consequently, V_o will decrease because $V_o = g_m V R_L$. It is the increase in reactance of C_C at low frequencies that controls and explains the reduction of V_o at low frequencies.

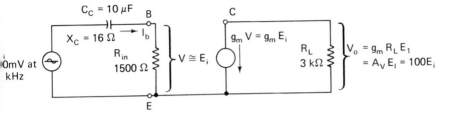

(a) Signal voltage levels at 1 kHz for Fig 16-1

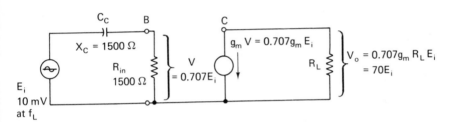

(b) Signal voltage levels at f

Figure 16-3 Coupling capacitor C_C does not affect V_o at 1 khz in (a), but it reduces V_o at f_L in (b).

16-3.2 Evaluating the Lower Cutoff Frequency. When the reactance of C_C has a magnitude equal to R_{in}, BJT input voltage, V, in Fig. 16-3(b) will be reduced to $0.707E_i$. This statement is proved in the following example.

Example 16-4: Prove that the magnitude of V equals $0.707E_i$ when $X_C = R_{in}$. Assume that frequency $f = f_L$ when $X_C = R_{in}$.
Solution: Since E_i divides between C_C and R_{in}, express V in general terms from the voltage divider relationship,

$$V = \frac{R_{in}}{R_{in} - jX_C} E_i$$

Substitute for $X_C = R_{in}$:

$$V = \frac{R_{in}}{R_{in} - jR_{in}} E_i$$

Cancel R_{in} in numerator and denominator:

$$V = \frac{E_i}{1 - j1}$$

Since $1 - j1 = \sqrt{1^2 + 1^2} \, /\underline{\tan^{-1}(1/1)} = \sqrt{2} \, \underline{/-45°}$

$$V = \frac{E_i}{\sqrt{2} \, \underline{/-45°}}$$

But $1/\sqrt{2} = 0.707$ and, neglecting the phase angle,

$$|V| = 0.707|E_i|$$

where $|\quad|$ means "magnitude only."

We conclude from Example 16-4 that when $X_C = R_{in}$, V will be reduced to $0.707E_i$, and consequently V_o will be reduced by 0.707 to

$$V_o = g_m V R_L = g_m R_L (0.707E_i) = 100(0.707E_i) = 70E_i$$

f_L is the frequency where $X_C = R_{in}$. This relationship shows how to evaluate f_L, or

$$X_C = \frac{1}{2\pi f_L C_C} = R_{in} \tag{16-6a}$$

Solving for f_L,

$$f_L = \frac{1}{2\pi C_C R_{in}} = \frac{0.159}{C_C R_{in}} \tag{16-6b}$$

Example 16-5: Evaluate f_L for Fig. 16-3.
Solution: From Eq. (16-6b),

$$f_L = \frac{0.159}{(10 \times 10^{-6})1500} = 10 \text{ Hz}$$

Example 16-6: Choose C_C for a cutoff frequency of 60 Hz.
Solution: Rewrite Eq. (16-6a):

$$C_C = \frac{0.159}{f_L(R_{in})} = \frac{0.159}{60(1500)} \simeq 1.7 \ \mu\text{F}$$

We see from a comparison of Examples 16-5 and 16-6 that *reducing C_C by a factor of 6 increases f_L by the same factor.*

16-3.3 Conclusion. A very useful conclusion may be drawn from Eq. (16-6a). The lower cutoff frequency is determined by a capacitor and the total resistance across the capacitor's terminals. For Fig. 16-3, C_C was the capacitor and R_{in} was the total resistance. By equating resistance to capacitive reactance we obtain an expression for f_L. *This principle will be used again in other circuits to express cutoff frequency.*

16-4 Low-Frequency Cutoff Due to Emitter Bypass Capacitor

16-4.1 Gain Reduction by the Bypass Capacitor. Assume for the time being that C_C is so large in Fig. 16-4(a) that reactance of C_C is negligible at very low frequencies, even down to 1 Hz. This assumption will allow us to concentrate on how the emitter resistor bypass capacitor C_E can change voltage gain at low frequencies.

When E_i has a frequency of 1000 Hz, reactance of C_E should be negligible and R_E will be effectively short-circuited. Voltage gain is then given by:

$$\text{at 1 kHz:} \quad A_v = \frac{V_o}{E_i} = g_m R_L = \frac{1}{30} \times 3000 = 100$$

At very low frequencies, reactance of C_E will be very large, much larger in fact than R_E. We conclude that as frequency is lowered, C_E changes from a

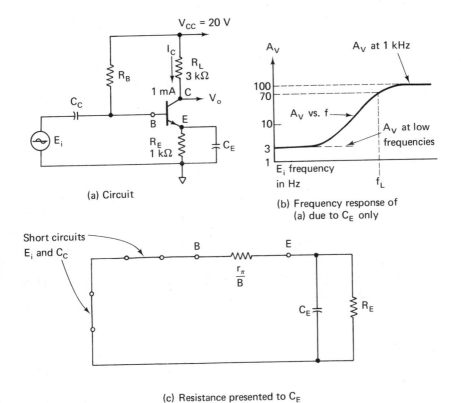

(a) Circuit

(b) Frequency response of (a) due to C_E only

(c) Resistance presented to C_E

Figure 16-4 Circuit to analyze effect of C_E on lower-frequency limit.

short circuit to an open circuit around R_E. So in some low-frequency range, voltage gain is

at very low frequencies: $\quad A_v = \dfrac{V_o}{E_i} = \dfrac{R_L}{R_E} = \dfrac{3 \text{ k}\Omega}{1 \text{ k}\Omega} = 3$

As shown in Fig. 16-4(b), A_v must change from 100 to 3. At point L, lower cutoff frequency is located when gain A_v (or V_o) is reduced by a factor of 0.707.

16-4.2 Evaluation of the Lower Cutoff Frequency. We use the conclusion of Section 16-3.3 to learn how C_E controls lower cutoff frequency f_L. First, find the total resistance appearing across C_E from Fig. 16-4(a). We see that R_E is across the terminals of C_E. But the BJT's emitter is also connected to one terminal of C_E; therefore, we must determine the equivalent resistance seen looking into the emitter terminal. Replace C_C by a short circuit and replace E_i by its internal resistance, which is zero ohms. The resulting circuit is modeled in Fig. 16-4(c) to show how base-leg resistance is divided by β when viewed from the emitter. The resulting resistance R presented to C_E is

$$R = R_E \left\| \frac{r_\pi}{\beta} \right. \tag{16-7}$$

Usually R_E is very large with respect to $r_\pi | \beta$, so $R \simeq r_\pi | \beta$. Now the reactance of C_E is $1/2\pi f_L C_E$ at f_L and equals R or

$$\frac{1}{2\pi f_L C_E} = R \simeq \frac{r_\pi}{\beta} \tag{16-8a}$$

Solving for f_L,

$$f_L = \frac{1}{2\pi (r_\pi/\beta)C_E} = \frac{0.159}{C_E(r_\pi/\beta)} \tag{16-8b}$$

Example 16-7: If $\beta = 50$ and $r_\pi = 1500 \ \Omega$ in Fig. 16-4(a), find the value of C_E required for a cutoff frequency of 100 Hz.
Solution: Evaluate $r_\pi/\beta = 1500/50 = 30 \ \Omega$; therefore, $R = (1000 \| 30) \ \Omega \simeq 30 \ \Omega$. From Eq. (16-8b), solve for C_E:

$$C_E = \frac{0.159}{f_L(r_\pi/\beta)} = \frac{0.159}{100 \times 30} = 0.053 \times 10^{-3} \text{ F} \simeq 50 \ \mu\text{F}$$

Note that r_π/β is much smaller than R_E.

We learn from Example 16-7 that large values are to be expected for bypass capacitors. Low-voltage electrolytic capacitors are available at reasonable cost for purposes of bypassing. We next consider how to make true our assumption that C_C has negligible reactance at the lower cutoff frequency determined by C_E.

16-5 Low-Frequency Cutoff with Both Coupling and Bypass Capacitors

When two or more capacitors are present in a circuit, it is usually *not* good practice for each to try and set the same cutoff frequency. It is possible for them to interact and cause unwanted oscillations. It is preferable to arrange that only one capacitor establish f_L, and for all other capacitors to be chosen for a much lower cutoff frequency at $0.1f_L$.

When both emitter bypass and coupling capacitors are present in a circuit, it is more economical for C_E to establish cutoff frequency f_L and choose C_C to cut off at $0.1f_L$. Thus C_C acts as an effective short circuit at f_L and our assumption in Section 16-4.1 is valid.

Example 16-8: Select coupling and bypass capacitors in the circuit of Fig. 16-4(a) for cutoff at $f_L = 100$ Hz.
Solution: From Example 16-7, choose $C_E = 50$ μF. Let $f_L = 100/10 = 10$ Hz in Eq. (16-6b). For margin, assume that C_E shorts out R_E, so that $R_{in} = r_\pi = 1500$ Ω:

$$C_C = \frac{0.159}{(0.1f_L)R_{in}} = \frac{0.159}{10 \times 1500} \simeq 10 \ \mu F$$

16-6 BJT High-Frequency Limits

At high frequencies, coupling and bypass capacitors act as short circuits and do not affect high-frequency performance. However, two *internal* transistor capacitances do control high-frequency performance. As shown in the common emitter model of Fig. 16-5(a), capacitance of the base–emitter junction is modeled by capacitor $C_{b'e}$. Collector–base junction capacitance is modeled by capacitor C_o. Any signal current entering the base terminal must flow through a small part of the base semiconductor material before it encounters the base–emitter junction. Resistance of the base material is modeled by $r_{bb'}$. We did not include $r_{bb'}$ in earlier BJT models because $r_{bb'}$ is only significant at very high frequencies.

As the frequency of E_i is increased, $C_{b'e}$ shunts current away from r_π. Since only current through r_π is multiplied by β, βI_b, and consequently output voltage $V_o = \beta I_b R_L$, will decrease with increasing frequency. Capacitor C_o connects the large output voltage V_o to input voltage E_i. Thus, as E_i goes positive to pump current into r_π, V_o goes negative to pull current away from

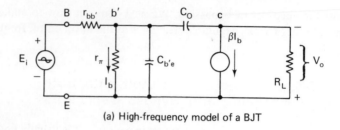

(a) High-frequency model of a BJT

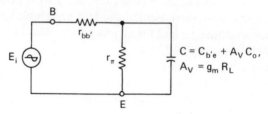

(b) Simplified input model of (a)

Figure 16-5 High-frequency BJT performance is analyzed from the high-frequencey model in (a) or simplified model in (b).

r_π. This action makes C_o act like a much larger capacitor *at the input*. This is called the *Miller effect*. Since current through C_o is increased because $V_o = A_v E_i$, C_o acts like a capacitor increased by voltage gain A_v. The effect of both C_o and $C_{b'e}$ is accounted for by an equivalent capacitor C in the simpler model of Fig. 16-5(b).

Values for C_o are given in data sheets and range from 2 to 20 pF. Other symbols used for C_o are C_μ, C_{oBo} or $C_{b'c}$. Values for $C_{b'e}$ may be given directly or, more commonly, they must be calculated from the data sheet electrical characteristic f_T. f_T is called *high-frequency current gain* or *current gain–bandwidth product*. Evaluate $C_{b'e}$ from

$$C_{b'e} = \frac{g_m}{2\pi f_T} - C_o \qquad (16\text{-}9)$$

Then evaluate equivalent input capacitance C from

$$C = C_{be} + A_v C_o \qquad \text{where } A_v = g_m R_L \qquad (16\text{-}10)$$

Example 16-9: The f_T of a BJT is specified as 150 MHz, and $C_o = 3$ pF at $I_C = 1$ mA. Evaluate $C_{b'e}$.
Solution: From Eq. (7-4b),

$$g_m = \frac{I_C(\text{mA})}{30} = \frac{1}{30} = 0.033$$

Applying Eq. (16-9),

$$C_{b'e} = \frac{0.033}{2\pi \times 150 \times 10^6} - 3 = 35\,\text{pF} - 3\,\text{pF} = 32\,\text{pF}$$

Example 16-10: The BJT of Example 16-9 is employed in the circuit of Fig. 16-5, where $r_{bb'} = 200\,\Omega$, $r_\pi = 2\,\text{k}\Omega$, and gain at 1 kHz of $A_v = 200$. (a) Evaluate f_H and (b) voltage gain at f_H.
Solution:

(a) From Fig. 16-5(b) and Eq. (16-10), $C = C_{b'e} + A_v C_o = 32\,\text{pF} + (200)(3\,\text{pF}) = 632\,\text{pF}$. Resistance presented to C is $R = r_{bb'} \| r_\pi = 200 \| 2000 \simeq 180\,\Omega$. The capacitive reactance of C is

$$X_C = \frac{1}{2\pi f C} = \frac{0.159}{fC} \tag{16-11}$$

and at $f = f_H$, $X_C = r_{bb'} \| r_\pi \simeq r_{bb'}$. Solving for f_H from Eq. (16-11) yields

$$f_H = \frac{0.159}{180 \times 632 \times 10^{-12}} = 1.4\,\text{MHz}$$

(b) At f_H, voltage gain is reduced to $0.707 \times A_v$, or A_v at $f_H = 0.707 \times 200 = 140$.

16-7 Temperature Limitations

16-7.1 Low-Temperature Limitation. In order to make connections between a circuit and the semiconductor materials of a BJT, it is necessary to have metal-to-semiconductor joints, as shown in Fig. 16-6. Metal and semi-

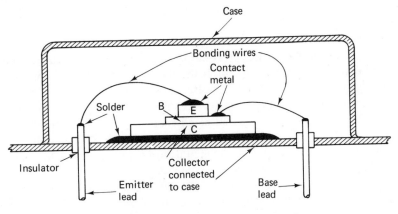

Figure 16-6 Simplified transistor construction.

conductor material contract at different rates when cooled. The joints are fabricated at high temperatures and undergo some stress when cooled to room temperature. To avoid joint fracture, a lower-temperature limit of about −65°C is imposed both for storage and operating temperature.

16-7.2. High-Temperature Limitation. Maximum temperature of a BJT is limited by the temperature, where solder will soften at locations shown in Fig. 16-6. This limit is typically 200°C for silicon transistors and 100°C for germanium transistors.

16-8 Power Limitation

16-8.1 Maximum Junction Temperature. When a BJT conducts collector current, I_C, and has a voltage drop, V_{CE}, between collector and emitter it develops power P_D, where

$$P_D = I_C V_{CE}$$

where I_C is in amperes, V_{CE} is in volts, and P_D is in watts. Power developed by the smaller base current is negligible. Just as a resistor is heated by passing a current, so is the BJT. Thus P_D will generate heat, mostly at the collector junction, as in Fig. 16-7(a). The temperature of the collector junction will rise until the rate at which P_D generates heat equals the rate at which heat is transferred to the environment or ambient. Ambient temperature means temperature of the air or fluid surrounding the BJT. The collector junction will stabilize at a temperature T_J. As noted in Section 16-7.2, the maximum temperature that can be allowed for a silicon transistor is about 200°C. Thus the *maximum junction temperature* that can be allowed would be symbolized by T_{Jmax} and for a silicon BJT, $T_{Jmax} = 200°C$. We conclude that *maximum power dissipation* P_{Dmax} that can be handled by a BJT will ultimately be determined by T_{Jmax}.

16-8.2 Thermal Resistance. It would be relatively easy to measure ambient temperature T_A for Fig. 16-7(a). If the BJT were mounted in a room, T_A would equal room temperature. The case temperature T_C should not be measured with a large bulb thermometer because it would cool the case upon contact. T_C could not be measured at all except by special techniques and equipment. Since it would be impractical to measure T_J and difficult to measure T_C, we must depend on the manufacturer to tell us how to estimate P_{Dmax} in terms of T_A and/or T_C.

Heat performance of the BJT is explained and simplified by one type of heat model in Fig. 16-7(b). The temperature rise from ambient T_A to junction

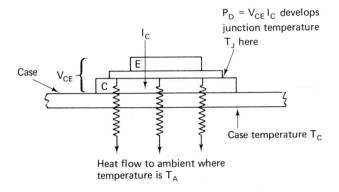

(a) BJT heat generation and flow

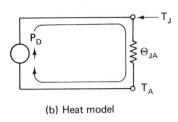

(b) Heat model

Figure 16-7 Heat generation and flow in (a) is modeled in (b).

T_J is explained by power P_D flowing through a *thermal resistance* θ_{JA}. Thermal resistance is the ratio of temperature difference, because of heat flow within the BJT, to power generated within the device. Mathematically we reason that ambient temperature T_A, plus temperature rise across θ_{JA}, or $P_D \times \theta_{JA}$, gives the junction temperature T_J. In equation form,

$$T_J = T_A + P_D \times \theta_{JA} \qquad (16\text{-}12a)$$

where T_J and T_A are in °C, P_D is in watts, and θ_{JA} has units of °C/W. *Maximum junction temperature* $T_{J\max}$ will occur when the BJT carries *maximum power* $P_{D\max}$, or

$$T_{J\max} = T_A + P_{D\max} \times \theta_{JA} \qquad (16\text{-}12b)$$

Example 16-11: $\theta_{JA} = 175°C/W$ and $T_{J\max} = 200°C$ for a BJT transistor. What is the maximum power that it can dissipate if (a) $T_A = 25°C$; (b) $T_A = 200°C$?

Solution:

(a) Rewrite Eq. (16-12b) as

$$P_{D\max} = \frac{T_{J\max} - T_A}{\theta_{JA}} = \frac{(200 - 25)°C}{175°C/W} = 1 \text{ W}$$

(b) $P_{D\max} = \dfrac{(200 - 200)°C}{175°C/W} = 0 \text{ W}$

Example 16-11 shows that power dissipation must be reduced or derated as ambient temperature is increased. This data may also be shown by the manufacturer as a *power-dissipation derating curve*, in Fig. 16-8. Here you enter the T_A axis at 25°C and read $P_{D\max} = 1$ W on the curve.

16-8.3 Heat Sink. A heat sink is a mass of metal attached to a BJT to remove heat faster so that the BJT can dissipate more power. A heat sink acts just like a radiator, in that it conducts heat. A room radiator conducts heat from circulating hot water to a room. A heat sink conducts heat from the BJT to the room. To see why a heat sink is needed, study the following example.

Example 16-12: It is necessary for a BJT to dissipate 2 W in an ambient temperature of 25°C. $T_{J\max} = 200°C$ and $\theta_{JA} = 175°C/W$ for this BJT. What temperature will the junction attain?
Solution: From Eq. (16-12a),

$$T_J = T_A + P_D \times \theta_{JA} = 25°C + 2 \text{ W} \times \frac{175°C}{W} = 375°C$$

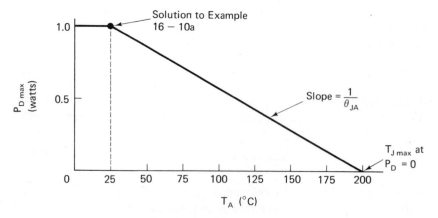

Figure 16-8 Power-dissipation derating curve.

Example 16-12 shows that 2W of power will heat the junction far above its maximum temperature and probably destroy it. However, the manufacturer also can specify thermal resistance from *junction to case*, θ_{JC}. For a BJT with $\theta_{JA} = 175°C/W$, the value of θ_{JC} is typically 35°C/W. The new BJT heat model is shown in Fig. 16-9(a). Often, *case-to-ambient* thermal resistance, θ_{CA}, is *not* given by a BJT manufacture if he gives θ_{JA}. This means that he intends for you to use the BJT with a heat sink.

Manufacturers of heat sinks specify *sink to ambient* thermal resistance, θ_{SA}. The smaller θ_{SA} is, the larger is the sink's heat-dissipation ability. The model for a heat sink is shown in Fig. 16-9(b).

The heat-sink manufacturer does not know what BJT you will use with his sink, and the BJT manufacturer does not know what sink you will use with his BJT. So each can only give data on his own product. When you select a BJT and heat sink they must be joined and heat must pass through the joint between BJT case and sink. The surfaces will not be absolutely flat and tiny insulating air pockets will be formed. To improve heat transfer from case to sink, an electrically insulating thermal grease should be applied sparingly between mating surfaces to fill the dead air space. θ_{CS} will then have values that range from 0.5 to 5°C/W.

From Fig. 16-9(c) we can now relate power dissipation and temperatures from

$$T_J = T_A + P_D(\theta_{SA} + \theta_{CS} + \theta_{JC}) \qquad (16\text{-}13a)$$

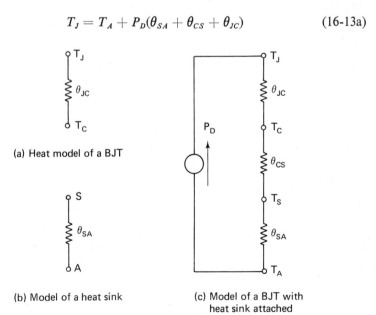

(a) Heat model of a BJT

(b) Model of a heat sink

(c) Model of a BJT with heat sink attached

Figure 16-9 BJT and heat-sink models.

Compare Eqs. (16-13a) and (16-12a) to see that

$$\theta_{JA} = \theta_{SA} + \theta_{CS} + \theta_{JC} \tag{16-13b}$$

Maximum T_J and P_D are related by

$$T_{Jmax} = T_A + P_{Dmax}(\theta_{SA} + \theta_{CS} + \theta_{JC}) \tag{16-13c}$$

Example 16-13: Given $P_D = 2$ W, $T_{Jmax} = 200°C$, $T_A = 25°C$, $\theta_{CS} = 1°C/W$, $\theta_{SA} = 10°C/W$, and $\theta_{JC} = 35°C/W$, what temperature will the junction attain?
Solution: From Eq. (16-13a),

$$T_J = 25°C + 2\ W(10 + 1 + 35)°C/W = 117°C$$

Compare the results of Examples 16-12 and 16-13 to see that installing the heat sink allows us to use the transistor without burning it up.

Problems

16-1 What is a frequency-response curve?

16-2 An amplifier's output voltage is 7 V rms at 1 kHz. What is the output voltage at f_H?

16-3 From tests obtained with the circuit of Fig. 16-1, $f_L = 100$ Hz and $f_H = 20,000$ Hz. Show that bandwidth BW is essentially equal to f_H.

16-4 An amplifier delivers 8 V rms to a 4-Ω speaker at 1 kHz. What power is developed in the speaker?

16-5 In Problem 16-4 speaker voltage is $8\ V \times 0.707 = 5.66$ V at f_L. What is the power developed at f_L?

16-6 What is the speaker power level in Problem 16-4 at f_H?

16-7 Find the changes in Example 16-1 if voltage gain is 50 at 1 kHz.

16-8 What circuit elements reduce amplifier output voltage at low frequencies?

16-9 What is the reactance of a 500-μF capacitor at 1 kHz?

16-10 In Fig. 16-3(a), $C_C = 1\ \mu$F and $R_{in} = 1500\ \Omega$. Evaluate f_L.

16-11 Compare Example 16-5 and Problem 16-10. Draw a conclusion regarding how f_L is changed when C_C is reduced by 10.

16-12 To improve (lower) low-frequency response, would C_C be increased or decreased in Fig. 16-3?

16-13 If $R_{in} = 10$ kΩ in Fig. 16-3, a capacitor of what size is required for $f_L = 100$ Hz?

16-14 In Example 16-7, what new value of C_E is required for $f_L = 10$ Hz?

16-15 Recalculate values of C_E and C_C in Example 16-8 for cutoff at $f_L = 200$ Hz.

16-16 $C_o = 5$ pF and $f_T = 100$ MHz for a BJT operated at $I_C = 1$ mA. Find $C_{b'e}$.

16-17 What changes result in Example 16-10 when a BJT is substituted with $C_{b'e} = 49$ pF and $C_o = 5$ pF?

16-18 A silicon BJT has a thermal resistance of $140°C/W$ and operates in an ambient temperature of $20°C$. What is the maximum power that it can dissipate?

16-19 In Example 16-11(a) the BJT is biased for $I_C = 0.2$ A. Find the maximum allowable V_{CE}.

16-20 Find P_{Dmax} if the BJT of Fig. 16-8 is operated in an ambient temperature of $112°C$.

16-21 A BJT with $\theta_{JA} = 175°C/W$ dissipates 0.5 W into an ambient temperature of $25°C$. Find its junction temperature.

16-22 A power BJT has $\theta_{JC} = 1.5°C/W$ and a heat sink has $\theta_{SA} = 5°C/W$. Assume that $\theta_{CS} = 1.5°C/W$ and find θ_{JA} for the resultant heat-sinked BJT.

16-23 A silicon BJT is heat-sinked for $\theta_{JA} = 8°C/W$. How much power can it dissipate in an ambient temperature of $25°C$?

17

Power Supplies

17-0 Introduction

A direct (dc) voltage is required to supply operating power to most electronic devices. Batteries are used for portable and low-current applications but are limited in operating time before they must be recharged (or replaced). Instead of a battery, we need a circuit arrangement that will convert the readily available ac voltage from a wall outlet to the desired dc supply voltage. As shown in Fig. 17-1, the circuit arrangement is called dc power supply.

Ideally a dc power supply should supply a dc voltage that (1) is all dc and contains no alternating component, and (2) does not change as more current is demanded from the supply. The most economical class of power supplies is the *unregulated* type. As will be shown in Sections 17-1 to 17-3, it has

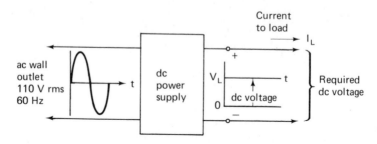

Figure 17-1 Dc power supply.

290

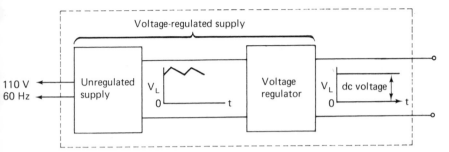

Figure 17-2 Adding a voltage regulator converts an unregulated power supply to a regulated power supply.

certain disadvantages. Its dc voltage decreases as current to the load increases. It may be difficult to obtain a particular dc voltage.

To minimize these disadvantages, a regulator section is added, as shown in Fig. 17-2. The resultant circuit is called a regulated power supply and will be studied in Sections 17-4 and 17-5. The regulator section allows us to obtain the desired dc voltage and practically eliminates changes in dc voltage with changing load current. Any ac voltage component in the unregulated supply is largely eliminated by the regulator. The only disadvantage with the regulated supply is its added cost.

17-1 Introduction to the Unregulated Supply

17-1.1 Introduction to the Full-Wave Bridge Rectifier. One of the most widely used unregulated types of power supply is the full-wave rectifier. As shown in Fig. 17-3(a), a transformer steps the ac line voltage down to the lower voltage levels required by most semiconductor devices. Four diodes are connected in a bridge arrangement that converts the transformer's secondary ac voltage to a pulsating dc voltage. Operation of diodes was covered in Fig. 3-7, where a signal generator is the source of alternating voltage.

Transformer voltages are specified in rms values. Rms secondary voltage E_{rms} is measured with an ac voltmeter. Peak secondary voltage E_m can be measured with a CRO. To convert easily from one measurement to the other, multiply E_{rms} by 1.4 to obtain E_m, or

$$E_m = 1.4E_{rms} \tag{17-1}$$

Example 17-1: The transformer of Fig. 17-3(a) is rated at 110 V to 24 V. What peak voltage would be measured across the secondary?
Solution: From Eq. (17-1), $E_m = 1.4 \times 24 \text{ V} = 34 \text{ V}$.

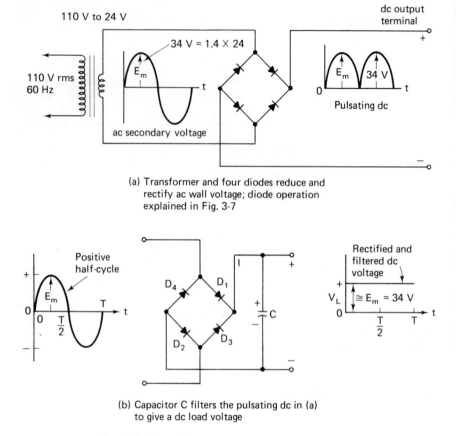

(a) Transformer and four diodes reduce and rectify ac wall voltage; diode operation explained in Fig. 3-7

(b) Capacitor C filters the pulsating dc in (a) to give a dc load voltage

Figure 17-3 Development of a full-wave bridge rectifier.

Equation (17-1) not only tells us the peak value of secondary ac voltage but also predicts the peak value of pulsating dc voltage developed at the output terminals.

17-1.2 Filter Capacitor. The pulsating dc voltage of Fig. 17-3(a) is not useful; it must be smoothed out. One means is by connecting a filter capacitor C across the dc output terminals in Fig. 17-3(b). Now capacitor C charges to a voltage V_L, equal to E_m. Actually there will be a slight voltage drop across the conducting diodes (see Fig. 3-7), but V_L essentially equals E_m.

No load is connected across C in Fig. 17-3(b). As far as the power supply is concerned, many electronic devices present a load that acts like a resistor. It is convenient to test power supplies by connecting a variable load resistor R_L that will draw load current from the supply. Then the effect of load current changes on the power supply's voltage can be readily observed.

As shown in Fig. 17-4(a), load R_L is wired across the dc terminals. The

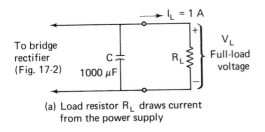

(a) Load resistor R_L draws current
from the power supply

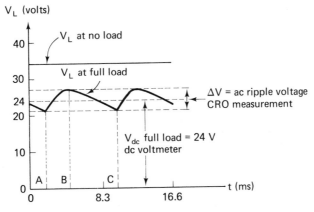

(b) Load voltage changes from 34 V at no load
to 24 V plus ripple at full load

Figure 17-4 Adding a load in (a) changes the load voltage, as in (b).

load voltage observed on a CRO would have the shape shown in Fig. 17-4(b). Assume that R_L is adjusted so that a full load dc current of $I_L = 1$ A is drawn from the supply. There are two significant differences between no-load and full-load voltages in Fig. 17-4(b). First, the full-load voltage has decreased, and second it now contains an alternating or *ripple* component.

At full load, the *average* voltage is measured with a dc meter and is called V_{dc}, to differentiate it from total instantaneous voltage, V_L. Peak-to-peak ac *ripple* voltage, ΔV, is measured on a CRO. This ripple voltage is caused by the discharge of current from capacitor C into load R_L during the time interval from B to C in Fig. 17-4(b). During time AB, the capacitor is recharged by current from the bridge rectifier. If a CRO is not available, an ordinary ac voltmeter may be connected across R_L to measure the approximate value of rms ripple voltage V_r. V_r will very nearly equal one-third of peak-to-peak ripple voltage ΔV, or

$$V_r \approx \frac{\Delta V}{3} \tag{17-2}$$

Example 17-2: Peak-to-peak voltage ΔV measures 5 V in Fig. 17-4. What reading would be obtained on an ac voltmeter connected across R_L?
Solution: From Eq. (17-2), $V_r = 5\ V/3 = 1.6\ V$.

The most direct way to analyze the way in which supply voltage depends on load current is to proceed on two separate paths. First, using dc meters only, investigate how dc load voltage V_{dc} changes with load current. Second, using only an ac voltmeter (or CRO), investigate how ac ripple voltage varies with dc load current. This separates dc from ac circuit operation and simplifies the analysis. Dc analysis will be accomplished in Section 17-2 and ac analysis in Section 17-3.

17-2 DC Voltage Regulation

17-2.1 Measurement of Voltage Regulation. Devices such as audio amplifiers, dc motors, and logic controls demand different currents from the power supply at different times. Normally we can determine minimum and maximum load-current requirements. For example, an audio amplifier would draw zero current when turned off and might draw 1 A at maximum volume. Thus loads are commonly specified in terms of minimum and maximum load currents. Unfortunately, the dc voltage delivered to the load, V_{dc} (also power-supply output voltage), changes as the load current changes.

The most convenient way to measure how the dc voltage of a power supply will vary with different load currents is to connect a variable load resistor to the supply terminals. Then measure dc load current I_L with a dc ammeter, and the dc load voltage V_{dc} with a dc voltmeter. This procedure is shown in Fig. 17-5(a). Both load voltage and current will contain an ac ripple component which does not effect the dc meter readings. The magnitude and control of this ac ripple will be discussed in Section 17-3.

As the load resistance R_L is reduced in Fig. 17-5, more dc load current I_L is drawn from the power supply. Dc load voltage V_{dc} becomes smaller as I_L increases (R_L decreases). For example, at no load, where $I_L = 0$, V_{dc} is equal to the peak value of secondary ac voltage, $E_m = 34$ V. V_{dc} is reduced to 24 V when I_L is increased to 1 A. These no-load and full-load operating points are shown at points O and A, respectively, in Fig. 17-5(a). The curve of V_{dc} versus I_L shows how dc load voltage varies with load current and is called the *voltage-regulation curve.* The no-load point O tells us the maximum dc power supply voltage at no load where $I_L = 0$. The full-load point A gives the lowest dc supply voltage at full load current. Data at these points will be useful in the next section and are summarized in Table 17-1.

17-2.2 Output Resistance. If a circuit is modeled as shown in Fig. 17-5(b), the load current and load voltage would vary in the same manner as the power supply. As more current was drawn through R_o, the voltage drop across R_o would increase and cause load voltage to decrease. This model is used to explain the power-supply *voltage-regulation* curve. The power supply has some net internal resistance caused by the diodes, transformer, capacitor, and resistance of wires connecting to the wall outlet. This net resistance is called power-supply *output resistance R_o*. It is difficult to predict, but simple to calculate, with test measurements of load voltage and current from the

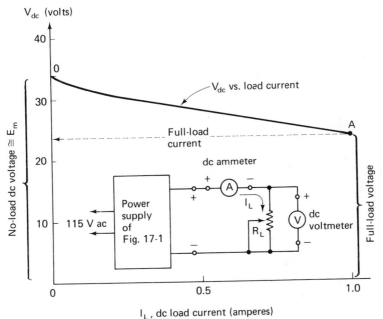

(a) Dc load voltage regulation

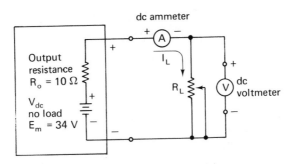

(b) Circuit model of the power supply

Figure 17-5 Voltage-regulation curve and dc model of a power supply.

Table 17-1 VOLTAGE-REGULATION DATA FOR FIG. 17-5

	No Load	Full Load
I_L	0	1 A
V_{dc}	34 V	24 V
R_L	∞ (open circuit)	24 Ω

equation

$$R_o = \frac{V_{dc\ no\text{-}load} - V_{dc\ full\text{-}load}}{I_{L\ full\text{-}load}} \qquad (17\text{-}3)$$

Example 17-3: Find output resistance of the power supply in Fig. 17-5.
Solution: From Table 17-1 and Eq. (17-3),

$$R_o = \frac{(34 - 24)\ \text{V}}{1\ \text{A}} = \frac{10\ \text{V}}{1\ \text{A}} = 10\ \Omega$$

Once the power-supply no-load dc voltage and output resistance are known, we can estimate the dc load voltage at any load current, from the dc loop equation of Fig. 17-5(b):

$$V_{dc} = V_{dc\ no\text{-}load} - I_L R_o \qquad (17\text{-}4)$$

Example 17-4: A transformer is employed in Fig. 17-5(a) with a secondary rms voltage of 6.3 V. From Eq. (17-1) it is found that $E_m = 1.41 \times 6.3$ V ≈ 9 V. Determine the dc load voltage V_{dc} at a load current of 0.5 A.
Solution: $V_{dc\ no\text{-}load} = 9$ V $= E_m$ and $I_L = 0.5$ A. In order to use Eq. (17-4), R_o must be known. There is no simple way to accurately predict R_o. Usually more expensive transformers are found to have $R_o \approx 2\ \Omega$, and inexpensive transformers may have an R_o greater than 25 Ω. Also, R_o is larger at small load currents ($\cong 10$ mA) than at higher currents. Normally R_o must be estimated, the circuit constructed and then R_o measured. Thus *assume* that $R_o = 5\Omega$. From Eq. (17-4),

$$V_{dc} = 9\ \text{V} - 0.5\ \text{A} \times 5\ \Omega = 9\ \text{V} - 2.5\ \text{V} = 6.5\ \text{V}$$

Example 17-5: The power-supply voltage of a portable transistor radio measures 9 V when off and 8 V when the radio is operating with full volume. The average load current is 0.1 A at full volume. What is the output resistance?
Solution: From Eq. (17-3),

$$R_o = \frac{9\ \text{V} - 8\ \text{V}}{0.1\ \text{A}} = 10\ \Omega$$

17-2.3 Need for Constant Voltage Regulation. Transformer–diode–capacitor power supplies have some disadvantages. Load voltage decreases with increasing load current. This is because there is no circuit element to regulate or hold load voltage constant. These supplies are classified as unregulated.

It may also be difficult or expensive to select the dc load voltage required at a particular current. This is because transformers come in standard sizes and the voltage needed may not be available except in a custom-made transformer. Even with a suitable transformer, the ac wall-outlet voltage can differ from location to location and will vary with main-line voltage changes.

These disadvantages—voltage selection and voltage dependence on load current—are overcome by voltage regulation. The simplest of these voltage-regulating circuits employs the Zener diode and will be discussed in Section 17-4.

17-3 How to Predict and Reduce AC Ripple Voltage

From Fig. 17-4 it is evident that ac ripple voltage increases as load current I_L increases. The worst case of ac ripple occurs at maximum load current and its peak-to-peak value, ΔV, can be predicted from the equation

$$\Delta V \approx 5 I_L \qquad \text{for } C = 1000 \ \mu F \qquad (17\text{-}5)$$

where I_L = dc load current in amperes, ΔV is in volts, and filter capacitor C equals 1000 μF. Equation (17-5) yields surprisingly accurate results considering its simplicity and is reasonable valid for load currents between 0.1 and 2 A.

Ac ripple voltage can be reduced by increasing capacitor C. If C is made twice as large, it stores twice as much charge. Thus capacitor voltage will discharge only half as much (half as much ripple) to furnish the same average dc load current between peaks. Figure 17-6 shows how the size of filter capacitor C can be selected to obtain a particular ripple voltage. For example, at a load current of 1 A, peak-to-peak ripple voltage will reduce from 5 V at 1000 μF to 2.5 V at 2000 μF and to 1 V at 5000 μF.

Peak-to-peak ac ripple voltage should be less than 10% of the dc load voltage, although this varies with the application and is not always true. For example, ripple magnitude is unimportant for a battery charger, but a 1% ripple or less may be required for a transistor-radio power supply.

Example 17-6: A power supply furnishes 0.5 A full-load current at 18 V and has a 500-μF capacitor. Calculate the approximate ac ripple voltage that will be measured on (a) a CRO and (b) an ac VTVM.

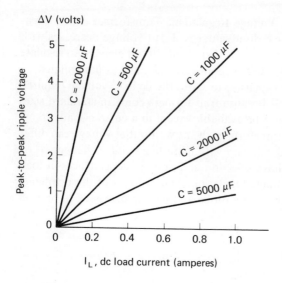

Figure 17-6 Filter-capacitor selection graph for full-wave rectifiers.

Solution: (a) From Fig. 17-6 enter the horizontal axis at 0.5 A and proceed vertically to the line $C = 500$ μF. Read $\Delta V = 5$ V from the vertical axis. (b) From Eq. (17-2), $V_r = 5$ V/3 ≈ 1.7 V.

Example 17-7: What are the maximum and minimum instantaneous values of load voltage for the supply of Example 17-6?
Solution: It is seen from Fig. 17-4(b) that ΔV is centered on V_{dc}. Thus

$$\text{maximum } V_L = V_{dc} + \frac{\Delta V}{2} = (18 + 2.5) \text{ V} = 20.5 \text{ V}$$

$$\text{minimum } V_L = V_{dc} - \frac{\Delta V}{2} = (18 - 2.5) \text{ V} = 15.5 \text{ V}$$

Example 17-8: If C is changed to 1000 μF, what is the effect on ripple voltage in Example 17-6?
Solution: At $I_L = 0.5$ A and $C = 1000$ μF in Fig. 17-6, read $\Delta V = 2.5$ V. Doubling C has reduced ac ripple by one-half.

Example 17-7 shows how to use the principle of superposition. First the dc problem is examined to find or specify the dc voltage at some load current. Then find the ac ripple voltage at the same load current. Finally, superimpose the ac ripple voltage on the dc voltage to predict the actual load voltage that contains both components. The Zener diode regulator will be studied in the next section to learn how it solves the problems of voltage selection and regulation.

17-4 Zener Diode Voltage Regulator

17-4.1 Zener Regulating Action. Section 3-4, on the characteristics and operation of the Zener diode, may be reviewed at this time if desired. It is found that when a Zener diode is reverse-biased, its terminal voltage remains reasonably constant, despite large current changes through the Zener. Connecting a load across the Zener results in *voltage regulation*. The value of constant Zener voltage is established in the manufacturing processes. Standard Zeners are available with voltage ratings between 2.4 and several hundred volts and with power ratings between $\frac{1}{4}$ and 50 W. If a 15-V supply is required, a 15-V Zener would be used.

To understand how the Zener diode regulates voltage, assume that a Zener is connected to the unregulated supply of Fig. 17-3. Next assume that a load resistor is connected to the Zener that will draw an assumed full-load current of $I_L = 0.5$ A. The resultant circuit is shown in Fig. 17-7(a). A series resistor R must be added to limit the Zener current to a small value in order to keep the Zener reverse-biased. Load voltage V_L will then equal Zener voltage $V_Z = 15$ V.

Maximum load power P_L is

$$P_L = I_L V_L = 0.5 \text{ A} \times 15 \text{ V} = 7.5 \text{ W} \qquad (17\text{-}6)$$

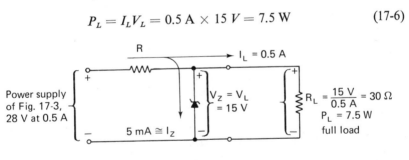

(a) Circuit currents at full load

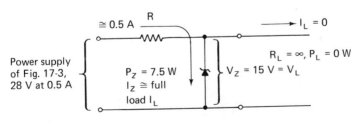

(b) Circuit currents at no load

Figure 17-7 Operation of a Zener voltage regulator.

If R_L is removed to simulate a no-load condition ($I_L = 0$), all the load current, and consequently load power, is absorbed by the Zener. This means that the Zener has to dissipate full-load power and must be selected and heat-sinked accordingly. Zener voltage changes very little, owing to its low Zener resistance during the increase in Zener current. Thus load voltage is regulated at about 15 V from no load to full load.

17-4.2 Selection of Zener Current Limit Resistor. Resistor R must be large enough to limit the Zener power to a maximum of full-load power as in Fig. 17-7(b). It also must be small enough to keep the Zener reverse-biased in Fig. 17-7(a). The limits of R were explored in Sections 3-4.4 and 3-4.5. In Zener power-supply applications, a simple design guide is used to limit the resistance R to a resistor value that is close to

$$R = \frac{\text{rectifier supply voltage} - \text{Zener voltage}}{\text{full-load current}} \times 0.8 \qquad (17\text{-}7)$$

where voltages are measured at full-load current. The factor 0.8 reduces the value of R to give the circuit a margin against the minimum instantaneous power-supply voltage (see Example 17-8).

Example 17-9: What value of R is required for the Zener regulator in Fig. 17-7?

Solution: From Fig. 17-7 and Eq. (17-7),

$$R = \frac{28 \text{ V} - 15 \text{ V}}{0.5 \text{ A}} \times 0.8 \approx 20 \text{ }\Omega$$

Example 17-10: The terminal voltage of a full-wave rectifier varies from 17 to 15.5 V as its output current varies from 0 to 0.25 A. A regulated 9-V power supply is needed to operate a transistor radio that draws a maximum of 0.25 A. What size (a) Zener and (b) Zener current limit resistor are required?

Solution:

(a) Select a 9-V Zener. The Zener power rating P_D is found from Eq. (17-6):

$$P_D = P_L = I_L V_L = 0.25 \text{ A} \times 9 \text{ V} = 2.25 \text{ W}$$

(b) From Eq. (17-7),

$$R = \frac{15.5 \text{ V} - 9 \text{ V}}{0.25 \text{ A}} \times 0.8 \approx 20 \text{ }\Omega$$

17-4.3 Zener Regulator Limitations. The Zener regulator provides a low-cost high-grade voltage regulator. If the full-load current will always be

drawn from the regulator, a simple Zener regulator would be the most economical. However, if the load is removed, the Zener would absorb the load
power and would burn out if it did not have a sufficient power rating.

High-powered Zeners are expensive; high-powered bipolar junction
transistors are much less expensive. A high-powered BJT and a low-powered
(low-cost) Zener may be combined to create an equivalent high-powered
Zener. This combination is called an *emitter-follower regulator* and will be
studied in Section 17-5.

17-5 Emitter-Follower Voltage Regulator

17-5.1 Theory and Operation. Load voltage is taken from the emitter of
the *npn* BJT in Fig. 17-8(a) and its base is connected to the Zener. This

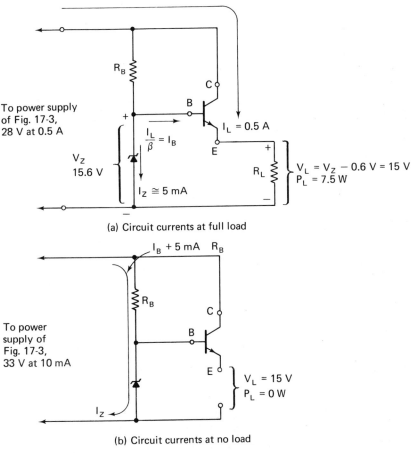

(a) Circuit currents at full load

(b) Circuit currents at no load

Figure 17-8 Operation of an emitter-follower regulator.

arrangement is designated a common-collector or emitter-follower circuit. Thus load voltage will "follow" or equal the Zener voltage. Actually, the load voltage will be slightly lower than V_Z, owing to the base–emitter voltage drop of the BJT. However, any error introduced by neglecting this voltage drop is about equal to the tolerance of the Zener voltage specification and can be neglected except in critical applications.

The Zener voltage and R_L determine load current. Transistor characteristics do not determine load current. The essential transistor characteristic is β because it determines the required value of base current I_B, where

$$I_B \approx \frac{I_L}{\beta} \tag{17-8}$$

Current through bias resistor R_B furnishes both base current drive I_B and Zener current. When the load is removed, as in Fig. 17-8(b), the Zener absorbs only I_B, *not* full-load current I_L. Since typically $\beta = 50$, the Zener absorbs only $\frac{1}{50}$ of maximum load current and consequently only $\frac{1}{50}$ of maximum load power. Thus low-powered, low-cost Zeners can be used to control much more power via the power transistor.

17-5.2 Bias-Resistor Selection. The bias resistor R_B must be small enough to furnish the required full-load base-current drive. This is usually the essential limit on R_B since Zener power dissipation is normally no problem, even for $\frac{1}{2}$-W Zeners. Thus R_B is selected from Fig. 17-8 using the following equation:

$$R_B = \frac{\text{full-load rectifier voltage} - \text{Zener voltage}}{\text{full-load current}/\beta} \times 0.8 \tag{17-9}$$

Example 17-11: What value of R_B is required in Fig. 17-8 if $\beta = 50$?
Solution: From Eq. (17-9),

$$R_B = \frac{28\text{ V} - 15.6\text{ V}}{0.5\text{ A}/50} \times 0.8 \approx 1\text{ k}\Omega$$

17-5.3 Practical Considerations. Zeners are inherently noisy because of avalanche breakdown. It is sound practice to connect a 10-μF capacitor across the Zener to shunt this noise away from the load. This capacitor also helps the Zener to stay in regulation during any sharp dips in the rectifier output voltage and should be added both to the Zener and to the emitter-follower regulator.

The transistor should have a power rating equal to the difference between power supplied by the rectifier and power delivered to the load at maximum load current. This calculation is made by multiplying maximum load current I_L by the quantity full-load rectifier voltage minus load voltage.

Example 17-12: What power rating is required for the BJT in Fig. 17-8(a)?
Solution: The difference between a full-load rectifier voltage of 28 V and a
load voltage of 15 V is 13 V. Thus $P_D = 13$ V $\times$ 0.5 A $= 6.5$ W for the BJT.
Note that Zener power is 10 mA $\times$ 15 V $= 150$ mW.

17-6 Bipolar or Positive and Negative Power Supplies

17-6.1 Center-Tapped Transformer. The power supplies studied thus far
have only two terminals. The negative terminal can be common (or grounded)
and the positive terminal used for a positive voltage. Alternatively, the positive terminal can be common (or grounded) and the negative terminal used
for a negative voltage. Many electronic devices need *both* (+) and (−) voltages with respect to a third common (or grounded) terminal. To obtain two
voltages we need either two secondary transformer windings or one center-
tapped secondary winding.

A transformer rated at 110 V to 24 V CT is shown in Fig. 17-9. The rating
voltages are specified in rms values. To understand the instantaneous voltages
as would be seen with a CRO, refer to the positive ac half-cycle in Fig. 17-9(a).
At the peak, terminal 1 is 12 $\times$ 1.4 $= 16.8$ V positive with respect to the

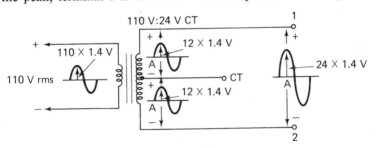

(a) Peak voltages for the positive half-cycle

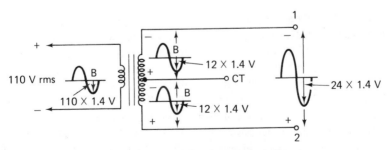

(b) Peak voltages for the negative half-cycle

Figure 17-9 Operation of a center-tapped transformer.

center tap, CT. Terminal 2 is 16.8 V negative with respect to CT. Terminal 1 is $24 \times 1.4 = 33.6$ V positive with respect to terminal 2. For the negative half-cycle in Fig. 17-9(b), the voltage polarities are reversed.

17-6.2 Full-Wave Center-Tapped Rectifier. In Fig. 17-10, two diodes, D_1 and D_2, are connected to terminal 1 of the transformer. Diodes D_3 and D_4 are connected to terminal 2. This converts alternating-current voltages to direct current. Also, as shown in Fig. 17-10, two filter capacitors are added to filter the pulsating dc voltages. Filter $C+$ smooths the positive dc voltage and

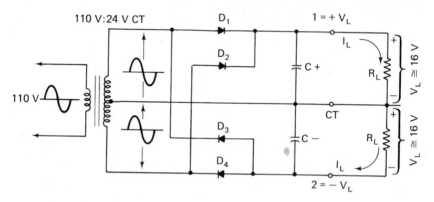

Figure 17-10 Four diodes and two capacitors convert the center-tapped transformer of Fig. 17-9 into a positive and negative power supply.

$C-$ smooths the negative dc voltage. D_1 and D_4 conduct on positive half-cycles, and D_2 and D_3 conduct on negative half-cycles. The voltage wave forms, ac ripple and dc regulation, operate the same as with the bridge-type rectifier. Terminal 2 furnishes a negative voltage with respect to CT, and terminal 1 furnishes a positive voltage with respect to CT. Thus this circuit arrangement is called a *bipolar* (positive and negative voltage) power supply.

17-6.3 Two-Value Supplies. A tapped transformer is more flexible than may be apparent. Normally, voltages are measured with respect to the center tap. As shown in Fig. 17-11(a), by treating the center tap as common, a positive supply voltage and a negative supply voltage may be obtained. Considering the common to be terminal 2 results in a two-value positive supply, as shown in Fig. 17-11(b). If terminal 1 in Fig. 17-11(c) is considered common, a power supply with two values of negative voltage is obtained.

By applying the principles of Sections 17-4 and 17-5, a regulator may be added across any pair of terminals in Figs. 17-10 and 17-11. The regulator will give the same advantages of voltage selection and regulation as it did with the bridge rectifier.

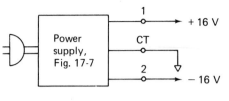

(a) Bipolar supply

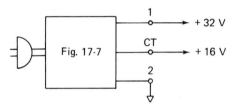

(b) Two-value positive supply

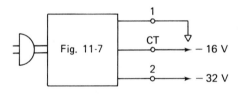

Figure 17-11 Power-supply connections for bipolar, positive, or
negative load voltages.

(c) Two-value negative supply

Problems

17-1 Why is a transformer generally found in a dc power supply?

17-2 Does dc voltage of an unregulated power supply decrease or increase when more current is drawn from it?

17-3 A transformer is rated at 110 V to 12.6 V. What is its peak secondary voltage?

17-4 An ac voltmeter reads 0.5 V across a dc power supply. What is the peak-to-peak ripple voltage?

17-5 If peak-to-peak ripple voltage is measured at 3 V by a CRO, what reading would be obtained on an ac voltmeter?

17-6 Sketch a test circuit to measure voltage regulation.

17-7 A transformer rated at 110 V to 12.6 V is used in an unregulated power supply. What is the no-load dc voltage?

17-8 Load voltage varies from 20 V to 17 V as load current is increased from 0 to 1 A. Find its output resistance, R_o.

17-9 What dc voltage is across a 1-A load if the power supply has a no-load voltage of 30 V and an internal resistance of 6 Ω?

17-10 Approximately what peak-to-peak ripple voltage will result in a full-wave rectifier supply of 0.2 A if the filter capacitor $C = 200$ μF?

17-11 In Problem 17-10, C is changed to 2000 μF. Find the resulting ripple voltage ΔV.

17-12 If a filter capacitor is doubled in size, is the ripple voltage doubled or halved?

17-13 What changes result in Example 17-6 if C is changed to 2000 μF?

17-14 What are the maximum and minimum instantaneous values of load voltage if $V_{dc} = 18$ V and $V_r = 0.83$ V?

17-15 In Fig. 17-7, $V_Z = 12$ V, $R_L = 24$ Ω, and $I_L = 0.5$ A at full load. Find the power absorbed by the Zener when the load is removed.

17-16 What maximum load current can be regulated by a 10-V Zener rated at 5 W?

17-17 In Fig. 17-7, $V_Z = 12$ V and I_L full load $= 0.5$ A. Find the required value of R.

17-18 For a full-load $I_L = 0.25$ A in Problem 17-17, what is the required value of R?

17-19 If $R = 10$ Ω in Fig. 17-7, find the full-load current.

17-20 In Fig. 17-8, $R_L = 30$ Ω and $\beta = 100$. Find the load current and the base current I_B.

17-21 What new value of R_B is required in Example 17-11 if $I_{Lmax} = 1.0$ A?

17-22 What power rating is required for the BJT in Problem 17-21? Assume that the unregulated supply voltage is 24 V at 1 A.

17-23 In Fig. 17-9, the transformer is rated at 110 V to 30 V CT. What no-load, bipolar dc voltages could be obtained with diodes and capacitors connected as in Fig. 17-10?

17-24 What other supply voltages can be obtained from the circuit of Problem 17-23?

Bibliography

CA3600E COS/MOS Transistor Array, File 619, RCA Solid State Division, Somerville, N.J., 1973.

Handling Precautions for MOS Integrated Circuits, Application Note ICAN 6000, RCA Solid State Division, Somerville, N.J., 1972.

COS/MOS Digital Integrated Circuits, SSD203A, RCA Corporation, Somerville, N.J., 1972.

COS/MOS Integrated Circuits Manual, Technical Series CMS-271, RCA Corporation, Somerville, N.J., 1972.

COUGHLIN, R. F. *Principles and Applications of Semiconductors and Circuits*, Prentice-Hall, Inc., Englewood Cliffs, N.J., 1972.

DRISCOLL, F. F. *Analysis of Electric Circuits*, Prentice-Hall, Inc., Englewood Cliffs, N.J., 1973.

DRISCOLL, F. F. and R. F. COUGHLIN *Solid State Devices and Applications*, Prentice-Hall, Inc., Englewood Cliffs, N.J., 1975.

KANE, J. The Field-Effect Transistor in Digital Applications, AN-219, Motorola Semiconductor Products, Inc., Phoenix, Ariz., 1970.

PETERSON, W. R. *Audio Applications of the RCA-HC1000 Hybrid Linear Power Amplifier*, AN-4474, RCA Solid State Division, Somerville, N.J., 1971.

———. *General Application Considerations for the RCA-HC1000 Hybrid Linear Power Amplifier*, AN4483, RCA Solid State Division, Somerville, N.J., 1971.

Sanken Technical Information, *Hybrid Power Amplifiers*, SI-1000 Series, 70–02 QA(2), Sanken Electric Co. ltd., Tokyo, 1972.

VILLANUCCI, R. S., A. W. AVTGIS, and W. F. MEGOW, *Electronic Techniques: Shop Practices and Construction*, Prentice-Hall, Inc., Englewood Cliffs, N.J., 1974.

Index